Gustavo Olinto da Silva

Theoretical study of TiO2 nanotubes and their interactions with gases

Gustavo Olinto da Silva

Theoretical study of TiO2 nanotubes and their interactions with gases

Interactions with CO and H2

ScienciaScripts

Imprint
Any brand names and product names mentioned in this book are subject to trademark, brand or patent protection and are trademarks or registered trademarks of their respective holders. The use of brand names, product names, common names, trade names, product descriptions etc. even without a particular marking in this work is in no way to be construed to mean that such names may be regarded as unrestricted in respect of trademark and brand protection legislation and could thus be used by anyone.

Cover image: www.ingimage.com

This book is a translation from the original published under ISBN 978-3-8417-1375-9.

Publisher:
Sciencia Scripts
is a trademark of
Dodo Books Indian Ocean Ltd. and OmniScriptum S.R.L publishing group

120 High Road, East Finchley, London, N2 9ED, United Kingdom
Str. Armeneasca 28/1, office 1, Chisinau MD-2012, Republic of Moldova, Europe
Printed at: see last page
ISBN: 978-620-8-28694-1

SUMMARY

I dedicate this work to the Lord, the almighty God, the I Am, who held my hand through the greatest difficulties. For only I know what I have been through on this journey. Fulfilling the Word written in Isaiah 41:13. I also dedicate this work to my beloved wife, for her patience, love, affection and understanding.

ACKNOWLEDGMENTS

To GOD, for all the provision, for having granted this work over my life. May all the honor and glory of this work be given to You. For everything came from You, for You. Nothing, exclusively nothing, escapes the control of Your hands;

To the Goiás State Research Support Foundation (FAPEG) for providing the master's degree scholarship;

To my beloved wife, for her strength, motivation, love, affection and for being a blessing in my life;

To my parents, for their encouragement, their concern and for wanting me to succeed so much;

To Prof. Dr. José Divino dos Santos, for his excellent teaching, patience and encouragement. As well as proving to be an excellent teacher and advisor, he was like a great friend, for his enormous help and understanding. I will always be grateful;

To Prof. Dr. Renato Rosseto for his willingness to participate in the committees to which this work was exposed and for the suggestions and contributions made at the beginning of the project and in the qualifying exam of this work;

To Prof. Dr. Leonardo Lopes da Costa for participating in the qualifying exam with his contributions and suggestions to the work;

To Prof. Dr. Joao Batista Lopes Martins for accepting the invitation to participate in the defense panel and for contributing to the completion of this work;

To the National Center for High Performance Processing of the University of Campinas (CENAPAD/UNICAMP-SP) and the Supercomputing Center of the Federal

University of Rio Grande do Sul (CESUP/UFRGS-RS), for providing computer access accounts to perform the calculations in this work;

To the bioanalytical manager Msc. Iram Mundim and the bioanalytical coordinator Msc. Sandro Gomes of the Institute of Pharmaceutical Sciences - ICF and to the industrial director of Clarion Biociências, Dr. Gladstone Santos de Souza for their support, allowing me to carry out this work during office hours;

To my leaders and beloved cell brothers for their friendship and for being my brothers in faith;

To my great friends who rejoice in my victories;

To my fellow master's students for the experiences we exchanged and for their encouragement;

"Because nothing is impossible for God"

(Lk 1:37)

"The wise will hear and grow in knowledge, and the learned will acquire wise counsel."

(Pr 1:5)

"The fear of the Lord is the beginning of knowledge; fools despise wisdom and instruction."

(Pr 1:7)

SUMMARY

The theoretical study of titanium dioxide nanotubes used a methodology for creating nanotube models from the unit cells of TiO_2 crystals (rutile and anatase) in specific planes, in this case anatase in the (101) plane and rutile in the (110) plane. The structures generated were subjected to quantum mechanics calculations using semi-empirical methods (PM7) and *ab-initio* DFT and HF methods, in the 6-31G and 6-311G bases. Models were created and it was found that the most stable structures were those with the largest diameters. Among the models that stood out in terms of energy variation were the anatase nanotube with the molecular formula $[[(TiO_2)_4]_{20}]_3$ and the rutile nanotube with the molecular formula $[[(TiO_2)_2]_{13}]_3$. By comparing models with the same number of atoms, it was found that the anatase nanotubes were more stable than the rutile nanotubes. In the *gap* evaluation using the semi-empirical and HF methods, the anatase nanotubes were classified as insulators due to the *gap* range between 6.00 eV and 10.00 eV, while the DFT values showed values between 2.00 eV and 2.50 eV, classifying them as semiconductors. The Ti-Ti and TiO bonds in the anatase models were found to be smaller than in the rutile models. Interactions of H_2 and CO with an anatase model were carried out and the adsorption of CO with the pure structure was verified. Adsorption with the H2 molecule was verified by doping tests with Pt and Au.

1 INTRODUCTION

Theoretical studies are making contributions to the development of explanations for various chemical systems using quantum mechanics. Quantum mechanics is able to generalize and understand particles and structures on a nanometric scale based on Schrodinger's equation, but due to mathematical limitations, it is necessary to use approximations to support experimental results (SZABO, OSTLUND, 1996).

The use of quantum mechanics is established in materials on an atomic scale, which behave differently from those observed on a macroscopic scale; quantum effects become more pronounced, they have new properties and the chemical elements exhibit different patterns of reactivity than structures on a larger scale (LEVINE, 2001).

Since the first scientific report on the synthesis of carbon nanotubes in 1991, published in the journal Nature by Sumio Iijima and collaborators (IIJIMA, 1991), countless studies on the synthesis, properties and applications of carbon nanotubes and inorganic nanotubes have been published.

The first scientific report on the synthesis of these nanotubes was in 1996 by researcher Patrick Hoyer, using the nanotube replication method (HOYER, 1996). Other methods have been developed that involve fewer steps and lower processing costs, such as hydrothermal synthesis and oxidative anodization. Today we know how to obtain ordered nanotubes with a higher degree of purity and how to dope them.

Titanium dioxide nanotubes are among the structures that are of interest due to their potential applicability, both experimentally and theoretically (GUO, *et al.* 2014).

Among the applications, one that stands out is the use of these structures as gas sensors, preventive instruments that inform of the appearance of gas leaks, avoiding possible destruction and death.

The theoretical study of nanotubes contributes to an understanding of their properties, trends towards greater stability, gaps and reactivity. It can help experimental synthesis studies and application studies of titanium dioxide nanotubes to achieve greater productivity and lower costs.

2. OBJECTIVES

- Evaluate the energy variation results and check the stability trends between the nanotube models obtained;
- Evaluate the energy variation results of anatase and rutile nanotubes and carry out a comparative study between rutile and anatase nanotubes;
- Check the winding energies in the nanotube formation reaction from the crystalline planes and determine the best formation trend;
- Compare the results of semi-empirical calculations (PM7) and *ab-initio* calculations (HF and DFT) on the 6-31G and 6-311G bases;
- Evaluate the *gap* results in semi-empirical (PM7) and ab-initio (HF and DFT) calculations on the 6-31G and 6-311G bases and classify rutile and anatase nanotubes according to the *gap* result (conductor, semi-conductor, insulator);
- Obtain the results of the distances and diameters of the nanotube models obtained;
- Compare the *gap* results obtained in the study with those obtained in the literature;
- Carry out the interaction between the titanium dioxide nanotube and the gases carbon monoxide (CO) and hydrogen gas (H_2) and compare the results obtained with data from the literature;

3. LITERATURE REVIEW

3.1. HISTORY AND APPLICATIONS OF NANOTUBES

In 1959, American physicist Richard Feynman, at a meeting of the American Physical Society, announced possibilities that until then could not be imagined, let alone realized. The possibility of being able to manipulate structures on an atomic scale (SANTOS, 2007).

In 1985, Kroto and collaborators achieved considerable results in their research into carbon structures, with the discovery of a new allotropic form of carbon called fullerenes (*buckballs*) after experiments on synthesis by laser evaporation of graphite. At first, these were structures of 60 and 70 carbon atoms arranged in a spherical shape (O' CONNELL, 2012), as can be seen in Figure 2.1. Since then, graphite and diamond have been the only known materials made up solely of carbons.

Figure 2.1 - Representation of C_{60}

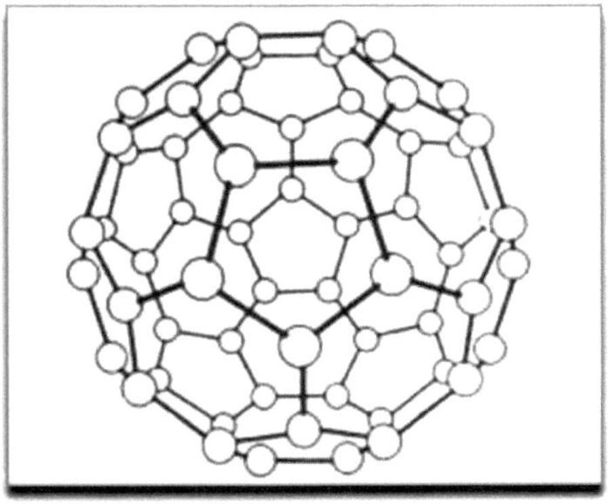

Source: HARRIS, 2009

In 1991, Sumio Iijima and colleagues succeeded for the first time in synthesizing carbon nanotubes using the arc discharge method. Iijima compared the cylindrical shapes visualized to needles containing 2 to 50 concentric tubes made up of graphene sheets. These concentric tubes with multiple layers were called *multiwall* nanotubes (IIJIMA, 1991). In 1993, Sumio Iijima and collaborators, as well as Donald Bethune and collaborators, succeeded in synthesizing carbon nanotubes with a single atom wall

(BETHUNE, *et al*. 1993; HARRIS, 2009).

In the 1980s and 1990s, it was realized that carbon could be organized in different ways, in addition to the forms that were already known (diamond, graphite). In the structure of diamond, the carbon atoms are linked to four other atoms, which are arranged in the form of a tetrahedron. The structure of graphite, on the other hand, is made up of sheets of graphene, flat structures of carbon, one atom thick, in which each carbon binds to three neighbors, resulting in a design similar to a set of hexagons. Graphite is formed by the overlapping of these sheets. In the structure of the best-known fullerene, C_{60}, the atomic bonds are organized in curved surfaces similar to a soccer. Nanotubes also result from the organization of carbon atoms into sheets, as in graphite, but instead of being stacked, each sheet is rolled into a cylinder (HAMMERS, 2003). Figure 2.2 illustrates the various allotropic forms of carbon.

Figure 2.2 - Allotropic forms of carbon

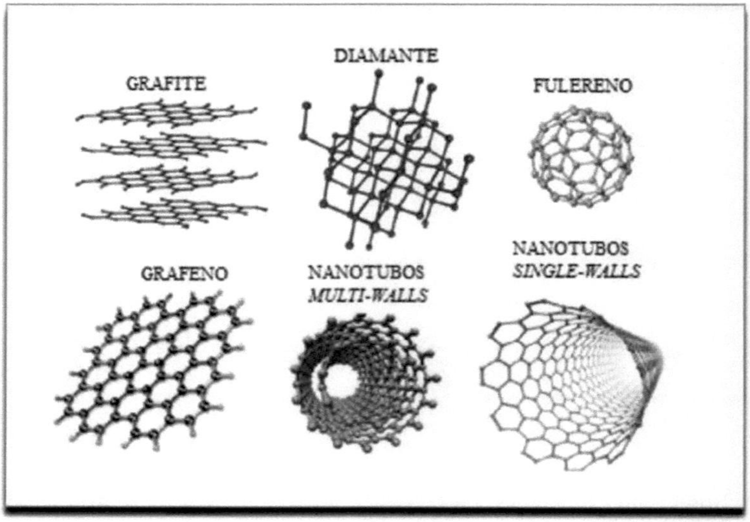

Source: ZARBIN, *et al* 2013 - Adapted.

From carbon nanotubes, a new range of possibilities opened up for the formation of new nanotubes made up of other chemical elements, with nanotubes of type X (made up of a single element), XY (made up of 2 types of atoms with a stoichiometry of 1:1) and XY2 (made up of 2 types of atoms with a stoichiometry of 1:2). In 1992, Tenne and his collaborators recorded research into nanotubes made of tungsten disulfide (WS_2), an XY2-type nanotube (TENNE, *et al.* 1992). Since then, research into inorganic nanotubes of the X, XY and XY2 types has expanded with compounds such as molybdenum disulfide (MoS_2) (FELDMAN, *et al.* 1995), silicon dioxide (SiO_2) (NAKAMURA, *et al.* 1995), nickel dichloride ($NiCl_2$) (HACOHEN, *et al.* 1998), gold (Au) (HUTLEEN, *et al.* 1998). The first records of synthesizing titanium dioxide nanotubes date back to 1996 with Patrick Hoyer (HOYER, 1996).

Since the discovery of these new allotropic forms of carbon, their properties, methods of obtaining them, purification methods and possible applications have been the subject of studies. Table 2.1 below shows some examples of studies into technological applications in different areas:

Table 2.1 - Studies and research into possible applications of nanotubes in different areas

Application	Objectives	References
Genetics - Chemical indicators	The use of nanotubes in microscopic analysis can trace DNA strands.	COLLINS & AUVORIS, 2000
Electronics - "Nanopigs"	Two nanotubes, connected to electrodes on a glass rod, can be opened and closed by changing the voltage.	COLLINS & AUVORIS, 2000
Medicine - Biomaterials	The use of nanotubes in the construction of *scaffolds,* a biomaterial used in bone regeneration.	GONÇALVES, 2012
Civil Engineering - Composites in cement	The use of carbon nanotubes to extend the useful life of concrete, the most widely used composite in construction worldwide.	LEITE, *et al.* 2014

Electronics - Lithium Batteries	Use of SnO nanotubes$_2$ as anodes in lithium batteries.	LIU et al, 2014

3.2. TITANIUM DIOXIDE NANOTUBES

3.2.1. Synthesis methods

After the synthesis of carbon nanotubes, other types of nanotubes began to be successfully synthesized, including titanium dioxide nanotubes. In 1996, Hoyer presented the first records of the synthesis of TiO nanotubes$_2$ using a process known as the **replication method** (HOYER, 1996).

In this process, a polymer mold suitable for forming titanium dioxide nanotubes was obtained, based on the naturally occurring structure of porous aluminum oxide. The tubular structure was formed by electrochemical deposition in the mold. After dissolving the polymer, titanium dioxide nanotubes were obtained and characterized. According to the author, the nanotubes obtained had a diameter of 70 to 100 nm (HOYER, 1996). The replication method involves several steps, making it difficult, expensive and generating only long nanotubes (SOUZA, F. M. 2011). Figure 2.3 is a microscopy of the titanium dioxide nanotubes synthesized by Hoyer.

Figure 2.3 - Scanning Electron Microscopy (SEM) of the synthesized TiO2 nanotubes.

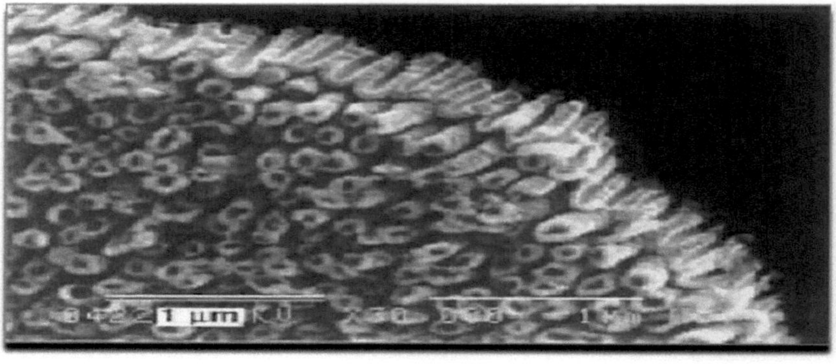

Source: (HOYER, 1996).

Two years after Hoyer's publication, Kasuga and colleagues succeeded in synthesizing TiO2 nanotubes from rutile in 1998. Using a solution of sodium hydroxide at a temperature of 110° C followed by washing with a solution of hydrochloric acid (KASUGA, *et al.* 1998). This synthesis carried out by Kasuga is called **hydrothermal synthesis**. The hydrothermal synthesis process, according to Bavykin, goes through the stages of exfoliation of nanometric multilayers, unbalance in the thickness of multilayers causing bending by mechanical tension (BAVYKIN, *et al.* 2004). This type of synthesis has been widely used by several researchers. In 2002, Yao *et al.* produced TiO2 nanotubes with a diameter of approximately 10 nm from anatase (YAO, *et al.* 2002). In 2004, Yuan and colleagues synthesized nanotubes of 8 to 10 nm in diameter and 10 to 100 nm in length (YUAN, *et. al.* 2004). According to Wu and collaborators in 2005, through hydrothermal synthesis they realized the formation of multilayer nanotubes (*multiwalls*) with a diameter of 8 to 12 nm and lengths of 200 to 1000 nm (WU, et al. 2005). Figure 2.4 shows titanium dioxide nanotubes obtained through hydrothermal synthesis.

Figure 2.4 - MET of titanium dioxide nanotubes obtained from hydrothermal synthesis: (a) Top view (b) Side view

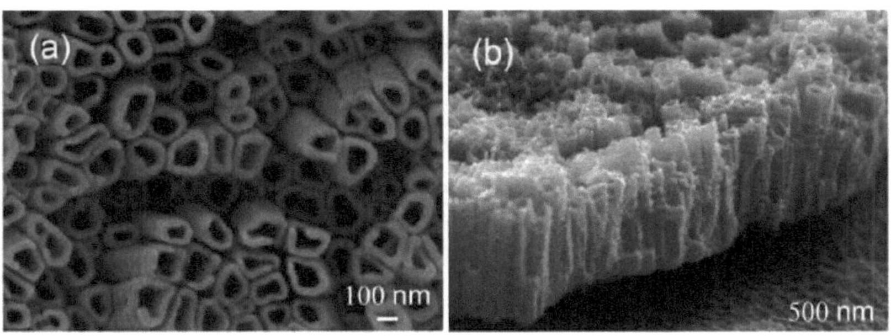

Source: (LU, et al. 2011) - Adapted.

In addition to the hydrothermal process, other synthesis routes are cited in research with the aim of forming TiO2 nanotubes, including the ***template* method** or also called the ***sol-gel* method** (KANG, *et al.* 2009) (MAIYALAGAN, *et al.* 2006). Kang *et al.* achieved the synthesis of well-aligned nanotubes with a diameter of 295 nm and a length of 6000 nm (Figure 3.8). Another method cited in the field is ***anodic oxidation***, cited by Macak et al. (MACAK *et al.* 2007), also used by Zhang et al. who succeeded in synthesizing TiO2 nanotubes with a diameter of approximately 100 nm, according to the author using fluoridic acid and phosphoric acid electrolytes (ZHANG, *et al.* 2008). Figure 2.5 shows nanotubes formed using the template method (sol-gel).

Figure 2.5 - SEM of TiO2 nanotubes after synthesis by the *template* method (sol-gel) (a) Lateral image, showing an open end of the TiO2 nanotubes (b) Top image of the TiO2 nanotubes

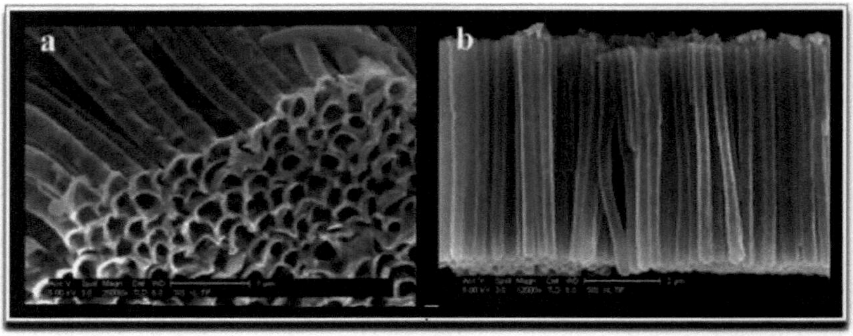

Source: (KANG, et al. 2009)

3.2.2. Theoretical Investigations

Various simulations, molecular modeling and theoretical investigations have been carried out on different types of materials (nanotubes, surfaces, crystals) formed by various compounds (TiO_2, ZnO, SnO_2, C, AsGa,) using quantum mechanics calculations, with the aim of finding out the values of energy parameters (ionization energy, variation in formation energy, gap) and comparing these results with experimental results.

Titanium dioxide nanotubes are among the structures that are of interest due to their potential applicability, both experimentally and theoretically (GUO*, et al.* 2014).

Using the *ab-initio* DFT+U method, Guo and colleagues demonstrated that nanotubes with oxygen vacancies in specific sites, which are common in titanium dioxide structures in crystal form, can show changes in the *gap* value of these structures. According to Guo, the *gap* value for an anatase-type nanotube in the (101) plane with a diameter of 8 m is 4.1 eV and with the simulation of vacancies, the calculations show that the *gap* value of the TiO2 nanotube is 0.8 eV (GUO, *et al.* 2014).

Nuzzi and Angellis carried out a theoretical study of TiO2 nanotubes based on DFT calculations on *single-walled* nanotubes from the anatase crystal in the [101] and [010] directions. With nanotubes ranging in diameter from 5 to 30 Â. Using the DFT method on an STO (*Slater Type Orbital*) basis, he found total energy values per TiO_2 unit, in which he found a low energy variation between the simulated models (<0.01 eV). Still using the DFT method, the B3LYP hybrid, in the 3-21G* base using the Gaussian 03 software (software used in the *ab-initio* calculations in this work), showed a *gap* value of 4.53 eV. In this same theoretical investigation, Nuzzi and Angellis also carried out the adsorption of TiO2 nanostructures with ferric acid (NUZZI & ANGELLIS, 2010).

Table 2.2 describes the characterization parameters of titanium dioxide nanotubes (crystal of origin of the nanotube, diameter, length, gap), which were found in research in the field of materials science (theoretical and/or experimental).

Table 2.2 - Data table of recent theoretical and experimental studies of titanium dioxide nanotubes

Data found in the literature						
	Type of study	**Type of TiO2 nanotube**	**Diameter (nm)**	**Compliment (nm)**	**gap (eV)**	**Reference**
1	Experimental	Anatase	60 á 80	770 -15210	3,20	FERNANDES, 2014
2	Experimental	Anatase	70	2000	3,20	LIU, *et al.* 2014
3	Experimental	Anatase	100-600	20000	-	MAJEWSKI, 2014
4	Experimental	Rutile	10 á 30	20000	-	LOSILLA, *et al.* 2014
5	Theoretical	Anatase	0,8	-	4,1 á 0,8	GUO, *et al.* 2014
6	Theoretical	Anatase	0,5 á 3	-	4,53	NUZZI & ANGELUS, 2014
7	Theoretical and	Anatase	-	-	3,14	YU, *et al.* 2014

	Experimental					
8	Theoretical and Experimental	Anatase	-	4000	3,06	SONG, *et al.* 2014
9	Theoretical and Experimental	Anatase	30*	220*	2,40 - 0,26	LI et *al.* 2014
10	Experimental	Anatase	25	-	-	ZHANG, *et al.* 2014

*Experimental Data

3.2.3. Experimental applications

The development of synthetic research is driven by the possibilities of using these structures in different areas of application. In medicine, TiO nanotubes$_2$ are prepared on a titanium surface by electrochemical anodization and having their walls coated with gold nanoparticles, they form a [TiO2-Au] compound. This compound is used to kill *Staphylococcus aureus* and *Escherichia coli* on surfaces in the absence of light (LI, *et al.* 2014). Or their use for bone reintegration by means of biomedical implants has been investigated. For this purpose, TiO2 nanotubes are generated using the anodic oxidation method (GANGULY & ZHAO, 2014).

In the field of molecular biology, membranes made of titanium dioxide are used for biofiltration of molecules, viruses and immune cells. According to the authors, membranes formed by nanotubes have shown interesting results in biofiltration, in experiments with glucose and insulin, showing high potential for application (SCHWEICHER & DESAI, 2014).

The use of nanotubes in energy applications has been investigated, such as their application in solar cells. In 2009, according to Schumik and colleagues, the increase in efficiency of solar cells made of titanium dioxide nanotubes was already considerable. In 2005, there were results of 0.04-0.05% efficiency, and in 2009, there was an efficiency of 4-5%, due to changes in the geometry of the solar cells, as well as the alignment of the nanotubes obtained with the optimization of the synthesis methods (SCHMUKI, *et al.* 2010). A "*micro-flower*" structure was developed from titanium dioxide nanotubes, which increased the efficiency of solar cells compared to a model formed by the normal arrangement of titanium dioxide nanotubes (KIM, *et al.* 2014). This structure is shown in Figure 2.6.

Figure 2.6 - Schematic representation of a "flower" solar cell made up of titanium dioxide nanotubes

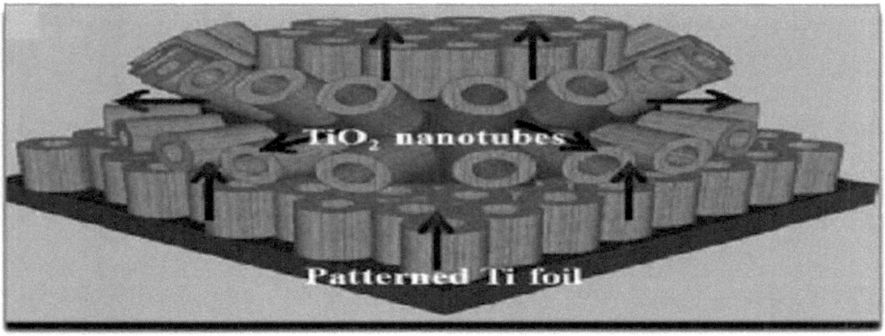

Source: (KIM, et al. 2014) - Adapted

In the area of photocatalysis, the titanium dioxide compound is already a highly applicable compound, and its nanotubes have demonstrated the same capacity, with a high potential for application in the environmental area. In 2009, titania nanotubes performed well in the photodegradation of dyes such as indigo carmine, as well as being easily recycled and reused in photodegradation reactions (COSTA, 2009). A recent study demonstrated the complete degradation of β-endosulfan (pesticide) in a photocatalytic process using titanium dioxide nanotubes doped with gold nanoparticles (THOMAS & CHITRA, 2014).

3.2.4. Application of titanium dioxide nanotubes as gas sensors

A chemical sensor can be defined as a device capable of detecting changes in the physical and/or chemical state of a system when exposed to a certain type of molecule. The chemical information is captured through a change in concentration or polarity, for example, in an electrical signal such as electrical conductivity, sensitivity, potential difference or resonance frequency (EGGINS, 2008).

Metal oxides are attractive for the manufacture of gas sensors due to factors such as low manufacturing costs, good thermal and chemical stability and production flexibility. Nanostructured metal oxides have good potential for use in the manufacture of gas sensors (BOLINK, *et al.* 2009).

Among nanostructured oxides, titanium dioxide nanotubes have attracted attention because of their physical and chemical properties (TIAN, *et al.* 2013). These tubular nanostructures have been considered one of the promising materials in the manufacture of gas detectors (ZHANG, *et al.* 2012).

Some factors in this process are being studied in order to optimize and increase the sensitivity of the titanium dioxide nanotube structure to gas detection. For example, increasing the surface area of the nanotubes and their morphology, as well as the ordering of the nanotubes, have proved to be of fundamental importance in assessing sensitivity to gases. This factor depends on the method used to synthesize the nanotubes. The use of doping has also been widely studied in order to improve sensor properties. Figure 2.7 shows the experimental results of using pure titanium dioxide nanotubes doped with niobium (Nb) as a sensor for some specific gases (GALSTYAN, *et al.* 2013).

Figure 2.7 - Response to gases (CO (500 ppm), NO2 (10 ppm), H2 (1 ppm), Ethanol (500 ppm) and Acetone (100 ppm)) in pure titanium dioxide nanotubes doped with niobium (Nb) at temperatures of 100° C, 200° C and 300 C.°

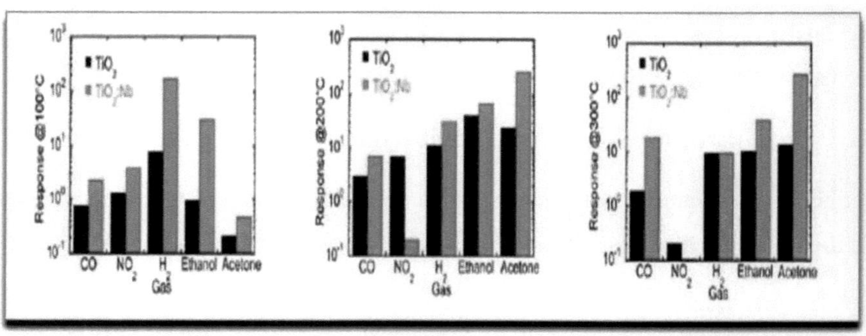

Source: (GALSTYAN, et al. 2012)

3.3. QUANTUM MECHANICS METHODS

3.3.1. Schrodinger equation

The Schrodinger equation was deduced in 1926 by the Austrian physicist Erwin Schrodinger. It was constructed on the basis of Borh's atomic model, incorporating Planck's quantum ideas, and is the fundamental equation of quantum mechanics. This equation can only be solved exactly for small systems such as the hydrogen atom; in polyatomic systems, some approximations must be made. Equation 2.0 describes the Schrodinger equation similarly to the energy equation in classical mechanics (SZABO, OSTLUND, 1996).

$$\frac{\hbar^2}{2m}\nabla^2\psi(r)+V(r)\psi(r)=E\,\psi(r) \qquad (2.0)$$

Equation 2.0 can be described by the representation of the Hamiltonian operator, describing the non-relativistic time-independent Schrodinger equation that describes the energy of quantum mechanical systems. Equation 2.1 where H represents the Hamiltonian operator, r and R represent the coordinates of the electrons and nuclei, respectively.

$$H\psi(r,R)=E\psi\,(r,R) \qquad (2.1)$$

Where the Hamiltonian operator can be expressed according to Equation 2.2:

$$H=-\sum_{i=1}^{N}\frac{1}{2}\nabla_i^2-\sum_{A=1}^{M}\frac{1}{2M_A}\nabla_A^2-\sum_{i=1}^{N}\sum_{A=1}^{M}\frac{Z_A}{r_{iA}}+\sum_{A=1}^{M}\sum_{B>A}^{M}\frac{Z_AZ_B}{R_{AB}}+\sum_{i=1}^{N}\sum_{j>i}^{N}\frac{1}{r_{ij}} \qquad (2.2)$$

The term RAB is the distance between the A and B nuclei, riA is the distance between the A nucleus and the i electron, rij is the distance between the i and j electrons, ZA and ZB are the atomic numbers of A and B and MA is the mass of the A nucleus in relation to the i electron. Equation 2.2 can be rewritten as Equation 2.3:

$$H=T_e+T_N+V_{Ne}+V_N+V_e \qquad (2.3)$$

The terms Te and TN are the electronic and nuclear kinetic energy operators respectively, the term VNE is the operator for the nucleon-electron attraction, VN is the operator for the nucleon-nucleon repulsive potential energy and V_e is the operator for the electron-electron repulsive potential energy. However, these equations do not make it possible to determine the quantum state of a polyatomic system, as mentioned above, and it is necessary to use approximations based on some chemical considerations represented by a mathematical formalism (SZABO, OSTLUND, 1996).

3.3.2. Born-Oppenheimer approximation

To solve polyatomic systems, an approximation applied to the Schrodinger equation is the Born Oppenheimer approximation, also known as perturbation theory, which involves solving Equation 2.2 in parts, separating the electronic and nuclear motions (LEVINE, 2001).

The perturbation theory uses the fact that the nucleus has a much greater mass than the electrons and as a consequence the electrons have a much greater speed than the nucleus. Based on these considerations, the nucleus behaves as if the electrons were a cloud of charge, while the electrons behave as if it were static, so that the electrons adapt instantaneously to changes in the spatial positions of the nucleus (LEVINE, 2001). Therefore, by reviewing Equation 2.3 we can cancel out the term T_N which represents the kinetic energy of the nucleus and consider the term VN which is the nucleus-nucleus repulsive energy as a constant.

With this we can rewrite Equation 2.3 and this equation can be separated into two other equations described in Equations 2.4 and 2.5.

$$H_{nuclear} = \sum_{A=1}^{M} \frac{1}{2MA} \nabla^2{}_A + \mathcal{E}_{total}(\{R_A\}) \tag{2.4}$$

$$\mathcal{E}_{total} = \mathcal{E}_{eletrônico} + \sum_{A=1}^{M} \sum_{B>A}^{M} \frac{ZAZB}{RAB} \tag{2.5}$$

Equation 2.4 is the Hamiltonian operator that deals with the contribution of nuclear energy to the system, while Equation 2.5 deals with the electronic contribution through the movement of electrons in the system. In this way we can represent the total energy of the system by summing the energies of the electronic and nuclear states (SZABO, OSTLUND, 1996). However, the Born-Oppenheimer approximation does not solve the quantum many-body problem, despite facilitating the solution of the Schrodinger equation, requiring other approximations.

3.3.3. Hartree-Fock theory

The Hartree-Fock theory was developed by D.R Hartree together with D.R V. Fock at the end of the 1930s in order to solve the Schrodinger equation for polyatomic systems. This methodology was the basis for the molecular orbital theory, stating that each electron can be described by a function (orbital), this movement being independent of the instantaneous movement of other electrons. The theory considers the use of the Born-Openheimer approximation for molecules and makes use of the Pauli Exclusion principle, which prohibits the presence of two electrons in the same quantum state (LEVINE, 2001).

Slater and Fock demonstrated that Hartree's method did not respect the principle of antisymmetry of the wave function and, by means of Slater's determinant, could satisfy the condition of this principle as well as the Pauli Exclusion principle. Using a system of two electrons with occupied spin orbitals χi and χj. If electron 1 is in χi and electron 2 is in χj or electron 1 is in χj and electron 2 in χi, it can be described as in Equation 2.6:

$$\Psi_{12}^{H}(x_1,x_2)=\chi_i(x_1)\chi_j(x_2) \qquad \Psi_{21}^{H}(x_1,x_2)=\chi_i(x_2)\chi_j(x_1) \tag{2.6}$$

Through a linear combination of these two products we can have a wave function that makes no distinction between the electrons and satisfies the Pauli exclusion principle in Equation 2.7:

$$\Psi(x_1,x_2)=\frac{1}{\sqrt{2}}\left(\chi_i(x_1)\chi_j(x_2)-\chi_j(x_1)\chi_i(x_2)\right)=\frac{1}{\sqrt{2}}\begin{vmatrix}\chi_i(x_1) & \chi_j(x_1)\\ \chi_i(x_2) & \chi_j(x_2)\end{vmatrix} \tag{2.7}$$

So for a system of n electrons using this same principle, we have a linear combination in a Slater determinant described in Equation 2.8:

$$\Psi(x_1, x_2, \ldots, x_N) = \frac{1}{\sqrt{N!}} \begin{vmatrix} \chi_i(x_1) & \chi_j(x_1) & \cdots & \chi_k(x_1) \\ \chi_i(x_2) & \chi_j(x_2) & \cdots & \chi_k(x_2) \\ \vdots & \vdots & \ddots & \vdots \\ \chi_i(x_N) & \chi_j(x_N) & \cdots & \chi_k(x_N) \end{vmatrix} \quad (2.8)$$

3.3.4. Hartree-Fock-Roothaan equation

For atoms, due to their spherical symmetry, the problem of solving the Hartree-Fock equation is simplified, but for molecules where this symmetry is lost, the method proposed by Roothaan is widely applicable. (CARVALHO, 2004) The Hartree-Fock-Roothaan equation can be expressed in matrix form according to Equation 2.9 (SANTOS, 2000).

$$\mathbf{FC} = \mathbf{SC}\,\varepsilon \quad (2.9)$$

In Equation 2.9, F represents the Fock matrix for the average field effect of all electrons on an electron in a given orbital. £ is a diagonal matrix, composed of £i elements, which is the energy of an electron in the orbital it occupies. S represents the orbital coverage matrix (SANTOS, 2000).

3.3.5. Semi-empirical methods

Semi-empirical methods are those which solve the Schrodinger equation from simple molecules to more complex systems, considering only the interactions of the atoms' valence electrons and using experimental parameters in order to find the values of the properties of these systems.

Compared to *ab-initio* calculations, more complex systems show a big difference in processing time, due to the greater speed of semi-empirical calculations, which, due to the information already available from the experimental data, mean that the calculations can be solved in a shorter space of time.

The pioneering work was carried out by Pople and collaborators. Dewar and collaborators developed a series of softwares so that semi-empirical calculations could be carried out more easily and in free software. The best known semi-empirical methods to date are AM1 (*Austin Model* 1) and MNDO (*Modified Neglect of Diatomic Overlap*) (DEWAR *et al*, 1985). These methods are used to simulate a series of experimental data such as equilibrium geometries, heats of formation, dipole moments and ionization energies. The latest parametrizations to have been developed are RM1 (*Reef Model 1*) (ROCHA *et al*, 2006), PM6 (*Parametric Method 6*) and PM7 (*Parametric Method 7*). With these methods, more than 70 elements from the periodic table have been parametrized, including the transition metals (STWEART, 2007).

3.3.6. *Ab-initio* methods

Ab-initio methods are based on mathematical models with the aim of solving the eigenfunctions of the Hamiltonian of the fundamental Schrodinger equation, unlike semi-empirical methods which use experimental parameters and consider only valence electronic interactions, but consider all the interactions involved in the polyatomic system (PARR *et al,* 1950).

The results are very reliable, but the computational time is high due to the large number of integrals to be solved. The main advantage of *ab-initio* methods is the precision with which the calculations are carried out.

3.3.7. Density Functional Theory - DFT

It is a quantum mechanical theory used to solve calculations of the properties of polyatomic systems (examples such as total energy, ionization energy) based on the determination of electronic density.

This is an alternative computational method to the ab-initio and semi-empirical methods first formulated in the work of Hohenberg, Kohn and Sham with self-consistent equations (SCF) based on experiments with electron gases (HOHENBERG, KOHN, 1964), (KOHN, SHAM, 1965).

Hohenberg and Kohn stated that the total electronic nuclear energy and all the ground state properties of a system are determined by the ground state electronic density. Without the need to calculate the wave function of N electrons (PARR, 1989).

4. METHODOLOGY

The research carried out simulations of small clusters of tubes formed with titanium dioxide from rutile and anatase crystals. First, the planes were obtained and, from the curling of the planes, the nanotubes. The whole process is carried out using algorithms created in the *shell bash* language on the Linux system (See Appendix A). The images and graphs of the structures and results found were made using the GNUPLOT, RASMOL and MOLDEN graphing software. (GNUPLOT, 2014; BERNSTEIN, 2014; MOLDEN, 2014).

4.1. PROPOSAL TO OBTAIN PLANES AND NANOTUBES FROM RUTILE CRYSTAL

In order to obtain the structures of the TiO2 pianos and nanotubes originating from the rutile crystal, we first started with the unit cell of the rutile crystal with its respective experimental lattice parameters (HOWARD, *et al.* 1991). The rutile crystal is a type of tetrahedral crystal system whose parameter values are:

$$\mathbf{a = b = 4.5937\ \text{Å}}$$

$$\mathbf{c = 2.9587\ \text{Å}}$$

$$\mathbf{(\alpha=\beta=\gamma= 90^o)}$$

Figure 3.1 illustrates the unit cell of the rutile crystal:

Figure 3.1 - *Ball-stick* representation of the unit cell of the rutile crystal (TiO2)

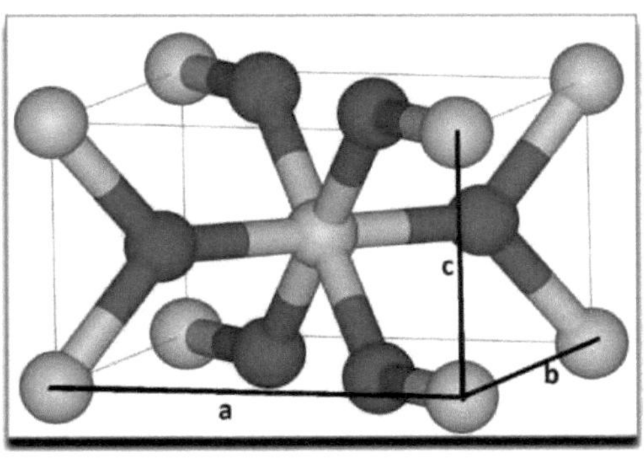

Source: (http://faculty.virginia.edu/metals/cases/sheehan3.html) - Adapted.

From these crystal values, we selected the (110) plane, and from the atoms contained in this plane, we grew the structure, generating the crystal planes. The (110) plane was chosen because it is a plane already known in the literature for its properties and potential applications in materials. According to Kowalski *et al.* the most thermodynamically stable crystal face of rutile titanium dioxide is the (110) face, which is dominant in its crystallites (KOWALSKI, *et al.* 2009). Other theoretical investigations of the rutile titanium dioxide compound have proven the better thermodynamic condition of the (110) plane. Ramamoorth and colleagues, using Vanderbilt pseudopotentials on plane wave bases in titanium dioxide surface structures in the (110), (100), (001) and (011) directions, compared the surface energies in the aforementioned directions, and found that the lowest surface energy was in the (110) direction (RAMAMOORTHY, *et al.* 1994). Beltran and collaborators reached similar results by density functional theory using the B3LYP functional (BELTRÀN, *et al.* 2008). Figure 3.2 shows the representation of the (110) plane and the atoms selected to form the basic unit of the structures.

Figure 3.2 - Plane (110) of the rutile crystal unit cell with the atoms selected to promote the growth of the planes and subsequently the nanotubes

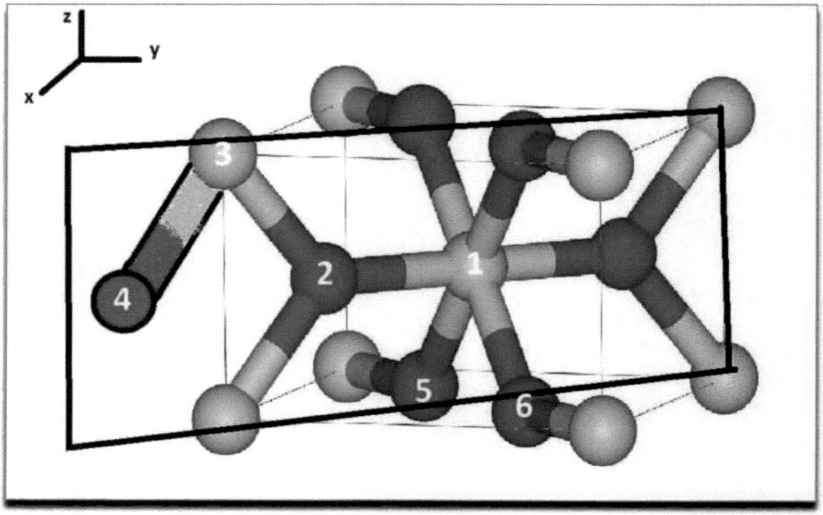

Source: (http://faculty.virginia.edu/metals/cases/sheehan3.html) - Adapted.

The atoms belonging to the (110) plane were selected, the atoms 1(Ti), 2(O), 3(Ti) and 4(O) in figure 5.2 and 2 more atoms outside this plane 5(O) and 6(O) to maintain the stoichiometry of the compound (1 Ti atom for 2 O atoms). So the basic structure of the planes and nanotubes generated from the rutile crystal in the (110) plane is 2 (TiO_2), shown in Figure 3.3:

Figure 3.3 - Basic unit of the planes and nanotubes originating from the rutile crystal - 2(TiO_2)

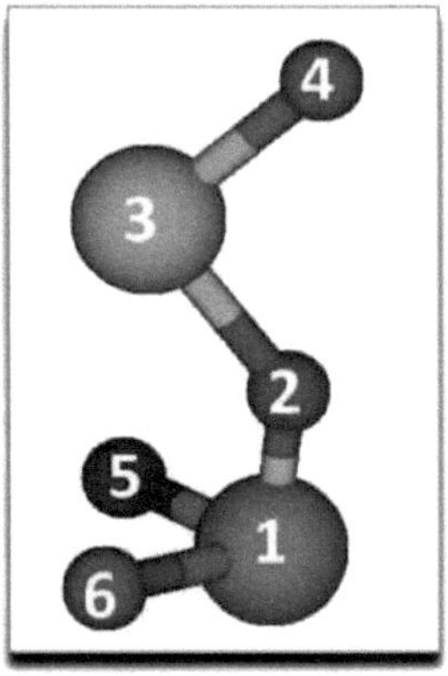

Source: Author

From this selected basic structure, the I(Ti) atom located in the center of the crystal was translated to the origin (0,0,0) of the Cartesian coordinate system (x,y,z). The same translation operation values were performed for all the other atoms. After the translation, the plane (110) was rotated 45° in relation to the z-axis in order to increase the length of the nanotubes around the x-axis and the height of the nanotubes on the z-axis. These operations were carried out in order to allow the structures generated to grow in the x-axis, which makes it easier to understand how the growth of planes and nanotubes occurs. In order to apply the translation and rotation operations to all the atoms in the structures generated, a general equation was described, which was used in the shell bash program (see Appendix A) to build the models of rutile planes and nanotubes.

First, the atom I(Ti) was translated to the point of origin with coordinates (0,0,0). In order to generalize the equation, we called these values Tx, Ty, Tz and applied a rotation of 45^0 , which we called φ.

Figure 3.4: Example of a rotation operation

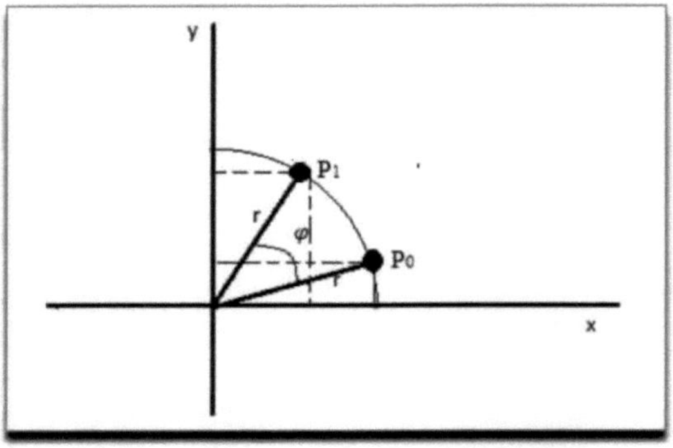

Source: Author

r - radius of the rotation circle;

φ - angle of rotation

Tx-Translation in the x-axis;

Ty-Translation in the y-axis;

Tz-Translation in the z-axis;

The starting position is given by

P_o - Starting position - (x, y, z);

$$x = r\cos\varphi;$$

$$y = r\,\mathrm{sen}\,\varphi;$$

$$z = z;$$

P_i - Position after rotation and translation - (xl, yl, zl);

Rotation about the z-axis occurs in the XY plane about the origin, where (x,y) is mapped to (x1, y1). The formula for applying this operation is given by:

$$x1 = x\cos\varphi - y\text{sen}\,\varphi$$

$$y1 = x\text{sen}\,\varphi + y\cos\varphi$$

Putting the translation operation and the 3-dimensional operation into matrix form, we have:

$$P1 = [x\ y\ z\ 1]\ x \begin{bmatrix} 1 & 0 & 0 & 0 \\ 0 & 1 & 0 & 0 \\ 0 & 0 & 0 & 0 \\ Tx & Ty & Tz & 1 \end{bmatrix} X \begin{bmatrix} \cos\varphi & \text{sen}\,\varphi & 0 & 0 \\ -sen\varphi & \cos\varphi & 0 & 0 \\ 0 & 0 & 0 & 0 \\ 0 & 0 & 0 & 0 \end{bmatrix}$$

Multiplying the matrix of initial coordinates by the translation matrix gives us a new matrix of initial positions that have already been translated.

$$P1 = [(x2 + Tx)\ \ (y2 + Ty)\ \ (y3 + Tz)\ \ 1]\ X \begin{bmatrix} \cos\varphi & \text{sen}\,\varphi & 0 & 0 \\ -sen\varphi & \cos\varphi & 0 & 0 \\ 0 & 0 & 0 & 0 \\ 0 & 0 & 0 & 0 \end{bmatrix}$$

The product of the translational coordinates and the rotation matrix, general equation (5.1), gives the translational and rotational coordinates (x1,y1,z1) for each atom.

$$P1 = [(x0 + Tx)\cos\varphi - (y0 + Ty)\,\text{sen}\,\varphi]\ \ [(x0 + Tx)\,\text{sen}\,\varphi + (y0 + Ty)\cos\varphi\ \ [z0 + Tz] \tag{3.1}$$

$$x1 = (x0 + Tx)\cos\varphi\ - (y0 + Ty)\,\text{sen}\,\varphi$$

$$y1 = (x0 + Tx)\,\text{sen}\,\varphi + (y0 + Ty)\cos\varphi$$

$$z1 = z0 + Tz$$

After defining the general equation (3.1) for the translation and rotation of the atoms in the rutile structures, the basic unit $(TiO_2)_2$ is repeated to obtain the planes and nanotubes that were analyzed in this study. Their growth depends on the number of $(TiO_2)_2$ units. First, the planes are created and grown along the z-axis, which describes their height, and along the x-axis, which describes their length. This plane is then wound along the x-axis, creating the nanotube. The rutile nanotube grows in length along the x-axis and in diameter in the yz-plane. The greater the number of $(TiO_2)_2$ units in the yz plane, the greater its diameter. The growth of the nanotube along the x-axis promotes the growth of the length levels of the nanotubes. Rutile nanotubes started to be generated from 6 units of $(TiO_2)_2$ in the YZ plane, because with only 5 units or less, there would be no stable nanotube formation.

4.2. PROPOSAL TO OBTAIN PLANES AND NANOTUBES FROM ANATASE CRYSTAL

In order to create the structures of the TiO pianos and nanotubes$_2$ originating from the anatase crystal, we first started with the unit cell of the anatase crystal with its respective experimental lattice parameters (HOWARD, *et al.* 1991). The anatase crystal, like rutile, is a tetrahedral crystalline system, but the arrangement of atoms and the amount of oxygen and titanium is greater than the unit cell of rutile. A representation of the anatase crystal unit cell is shown in Figure 3.5. The parameter values for the anatase crystal are:

$$\mathbf{a = b = 3.7845\ Å}$$

$$\mathbf{c = 9.5143\ Å}$$

$$\mathbf{(\alpha=\beta=\gamma= 90^o)}$$

Figure 3.5 - *Ball-stick* representation of the anatase crystal unit cell $(TiO\ )_2$

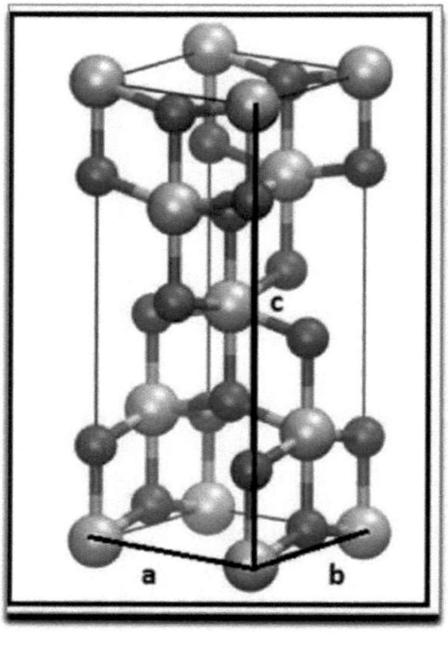

Source: (MOELLMANN, *et al.* 2012) - Adapted

Based on the crystal parameters, we selected the (101) plane equivalent to the (011) plane, and from the atoms contained in this plane, we grew the structure, generating the crystal planes. The (101) plane was chosen because it is a plane already known in the literature for its properties and potential applications in materials. Gong and colleagues carried out a detailed theoretical and experimental study of the (101) surface of the anatase crystal (using tunneling microscopy and spectroscopy and theoretical calculations using pseudopotentials) demonstrating the great potential for photocatalysis applications of the specified surface (GONG, *et al.* 2006). Previous studies have also confirmed the importance of the (101) plane of the anatase crystal. Diebold and co-workers carried out research with several planes and found a great similarity between the (101) plane of anatase and the (110) plane of rutile crystal, in terms of density and atomic surface structure (DIELBOLD, *et al.* 2003). Shklover and colleagues, using high-resolution transmission electron microscopy, found anatase nanocrystallites with (101) planes to be the dominant phase among other anatase crystals in studies of semiconductor surfaces (SHKLOVER, *et al.* 1997). Figure 3.6 shows the plane selected for nanotube growth.

Figure 3.6 - Plane (101) of the anatase crystal unit cell with the atoms selected to promote the growth of the planes and subsequently the nanotubes

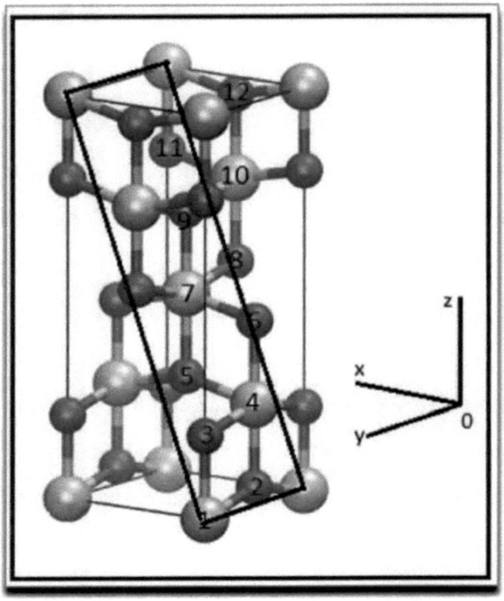

Source: (MOELLMANN, *et al.* 2012) - Adapted

Atoms belonging to the (101) plane were selected, atoms 1 to 12 (Figure 3.7). In the case of anatase, there was no need to use atoms outside the plane, as was the case with rutile. The basic structure of the planes and nanotubes generated from the anatase crystal is 4 units of (TiO_2). The basic unit of anatase nanotubes is illustrated in Figure 3.7:

Figure 3.7 - Basic unit of the planes and nanotubes originating from the anatase - 4(TiO_2) crystal

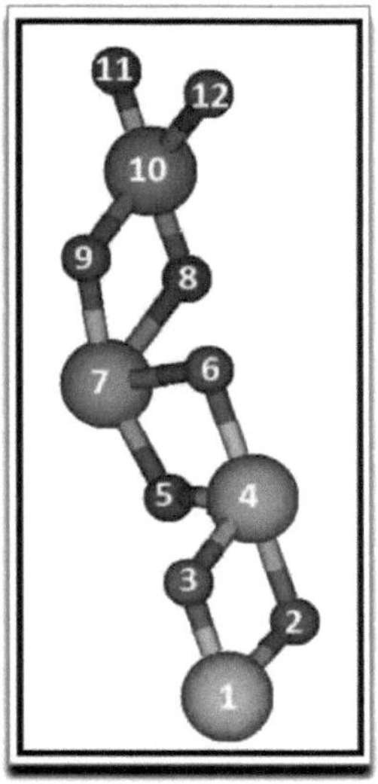

Source: Author

From this selected basic structure, the I(Ti) atom was translated to the origin (0,0,0) of the Cartesian coordinate system (x,y,z). The same translation operation values were performed for all the other atoms. After the translation, two rotations of the plane (101) were carried out, the first in relation to the x-axis and the second, a rotation of 90° in relation to the y-axis, in order to promote growth in the length of the nanotubes around the x-axis and growth in height on the z-axis. This procedure is the same as that used on the rutile crystal, which means that the planes and nanotubes of both crystals have the same growth shape. In order to apply the translation and rotation operations to all the atoms in the structures generated, the translation and 1^a rotation and 2^a rotation operations were carried out separately, unlike for rutile, where a single general equation was used. The operations are described in the *shell bash* program (see Appendix A) for building the models of pianos and anatase nanotubes.

After carrying out the operations for the atoms of the basic unit of anatase $(TiO_2)_4$, these units are repeated to obtain the pianos and nanotubes that were analyzed in this study. Their growth depends on the number of $(TiO_2)_4$ units. First, the pianos are created and grown along the z-axis, which describes their length, and along the x-axis, which describes their length. This piano is then wound along the x-axis, creating the nanotube. The anatase nanotube grows in length along the x-axis and in diameter along the yz-axis. The greater the number of $(TiO\ )_{24}$ units on the yz-piano, the greater its diameter. The growth of the nanotube along the x-axis promotes the growth of the nanotube length levels. Anatase nanotubes are now generated from 2 units of $(TiO_2)_4$, on the YZ piano, because with only 1 unit there would be no stable nanotube formation.

4.3. NANOTUBE NAMES

In order to identify the nanotubes, a notation system was adopted, containing information on the crystal that gave rise to the structure, the number of $(TiO_2)_2$ or $(TiO_2)_4$ units that form the curvature and diameter of the nanotube in the YZ plane, as well as the number of units of $(TiO\ )_{22}$ or $(TiO\ )_{24}$ that define the height or number of levels that the nanotube has along the nanotube's X axis and whether it is an optimized or non-optimized nanotube.

General nomenclature:

NT_ANT or **RUT** (symbol of the crystal of origin, ANT equivalent to anatase and RUT to rutile crystal)_n (Determines the number of units of $(TiO\ )_{22}$ or $(TiO\ )_{24}$ that form the diameter of the nanotube) _m (Determines the number of units of $(TiO2)_2$ or $(TiO_2)_4$ that form the length of the nanotube)**_ot (**Indicates whether the structure is optimized**)**

Example: **NT_RUT_10_03_ot** (Nanotube formed from rutile crystal, with 10 units of $(TiO_2)_2$ forming the diameter of the structure in the YZ plane and 03 units of $(TiO_2)_2$ forming the length or height in the X axis, optimized nanotube).

4.4. PLAN NAMES

To identify the pianos, a similar notation system was used, containing information on the crystal that gave rise to the structure, the number of $(TiO_2)_2$ or $(TiO_2)_4$ units that form the length of the plane, and the number of $(TiO\ )_{22}$ or $(TiO\ )_{24}$ units that define the height of the plane. Finally, whether the piano is optimized or non-optimized.

General nomenclature:

PL_ANT or **RUT** (symbol of the crystal of origin, ANT equivalent to anatase and RUT to rutile crystal)_n (Number of units of $(TiO_2)_2$ or $(TiO\ )_{24}$ that form the length of the plane)_m (Number of units of $(TiO_2)_2$ or $(TiO_2)_4$ that form the height of the plane)_ot (Indicating whether the structure is optimized)

Example: **PL_RUT_10_03_ot** (Plan formed from the rutile crystal, with 10 units of $(TiO_2)_2$ forming the length and 03 units of $(TiO_2)_4$ forming the height, optimized plan).

4.5. MOLECULAR FORMULA OF THE STRUCTURES

$[[(TiO\ )_{22}\ ]n]m$ - Molecular formula of the structures originating from the rutile crystal. Where n is the number of units $(TiO\ )_{22}$ in the YZ plane that form the curvature of the nanotubes and the length for the planes and m is the number of units $(TiO_2)_2$ in the X axis that form the length of the nanotubes and the height of the planes;

$[[(TiO_2)_4]n]m$ - Molecular formula of the structures originating from the anatase crystal; where n is the number of units $(TiO\ )_{24}$ in the YZ plane that form the curvature of the nanotubes and the length for the planes and m is the number of units $(TiO_2)_4$ in the X axis that form the length of the nanotubes and the height of the planes;

Table 3.1 shows the relationship between the names of the structures and their respective molecular formulas.

Table 3.1 - Table with the nomenclature and molecular formula of anatase nanotubes with 1 and 2 levels

Structure name	**n$(TiO_2)_4$**	**m$(TiO_2)_4$**	**Number of atoms**	**Molecular Formula**
NT-ANT-2-1-ot	2	1	24	[[(TiO2)4]2]1
NT-ANT-3-1-ot	3	1	36	[[(TiO2)4]3]1
NT-ANT-4-1-ot	4	1	48	[[(TiO2)4]4]1
NT-ANT-5-1-ot	5	1	60	[[(TiO2)4]5]1
NT-ANT-6-1-ot	6	1	72	[[(TiO2)4]6]1
NT-ANT-7-1-ot	7	1	84	[[(TiO2)4]7]1
NT-ANT-8-1-ot	8	1	96	[[(TiO2)4]8]1
NT-ANT-9-1-ot	9	1	108	[[(TiO2)4]9]1

NT-ANT-10-1-ot	10	1	120	[[(TiO2)4]10]1
NT-ANT-11-1-ot	11	1	132	[[(TiO2)4]11]1
NT-ANT-12-1-ot	12	1	144	[[(TiO2)4]12]1
NT-ANT-13-1-ot	13	1	156	[[(TiO2)4]13]1
NT-ANT-14-1-ot	14	1	168	[[(TiO2)4]14]1
NT-ANT-15-1-ot	15	1	180	[[(TiO2)4]15]1
NT-ANT-16-1-ot	16	1	192	[[(TiO2)4]16]1
NT-ANT-17-1-ot	17	1	204	[[(TiO2)4]17]1
NT-ANT-18-1-ot	18	1	216	[[(TiO2)4]18]1
NT-ANT-19-1-ot	19	1	228	[[(TiO2)4]19]1
NT-ANT-20-1-ot	20	1	240	[[(TiO2)4]20]1
NT-ANT-2-2-ot	2	2	48	[[(TiO2)4]2]2
NT-ANT-3-2-ot	3	2	72	[[(TiO2)4]3]2
NT-ANT-4-2-ot	4	2	96	[[(TiO2)4]4]2
NT-ANT-5-2-ot	5	2	120	[[(TiO2)4]5]2
NT-ANT-6-2-ot	6	2	144	[[(TiO2)4]6]2
NT-ANT-7-2-ot	7	2	168	[[(TiO2)4]7]2
NT-ANT-8-2-ot	8	2	192	[[(TiO2)4]8]2
NT-ANT-9-2-ot	9	2	216	[[(TiO2)4]9]2
NT-ANT-10-2-ot	10	2	240	[[(TiO2)4]10]2
NT-ANT-11-2-ot	11	2	264	[[(TiO2)4]11]2
NT-ANT-12-2-ot	12	2	288	[[(TiO2)4]12]2
NT-ANT-13-2-ot	13	2	312	[[(TiO2)4]13]2
NT-ANT-14-2-ot	14	2	336	[[(TiO2)4]14]2
NT-ANT-15-2-ot	15	2	360	[[(TiO2)4]15]2
NT-ANT-16-2-ot	16	2	384	[[(TiO2)4]16]2
NT-ANT-17-2-ot	17	2	408	[[(TiO2)4]17]2
NT-ANT-18-2-ot	18	2	432	[[(TiO2)4]18]2
NT-ANT-19-2-ot	19	2	456	[[(TiO2)4]19]2
NT-ANT-20-2-ot	20	2	480	[[(TiO2)4]20]2
NT-ANT-2-3-ot	2	3	72	[[(TiO2)4]2]3

4.6. OPTIMIZATION OF TITANIUM DIOXIDE PLANES AND NANOTUBES

The models obtained were first subjected to energy calculations using the semi-empirical method on the PM7 basis and ab-initio HF and DFT calculations with 6-31G and 6-311G basis functions. The MOPAC2012 software (STEWART, 2014) was used for the semi-empirical method and GAUSSIAN 03 (GAUSSIAN, 2014) for the ab-initio calculations. The quantum mechanics calculations performed by the software provide information such as total energy, heat of formation, dipoles, ionization potential, computational time, bond distances and homo-luminosity. To perform these calculations, processors were used at the National Center for High Performance Processing in Sao Paulo (CENAPAD-SP) at the State University of Campinas (UNICAMP-SP) and at the National Center for Supercomputing in Rio Grande do Sul (CESUP-RS) at the Federal University of Rio Grande do Sul (UFRGS-RS).

The models were generated from the experimental lattice parameters of the rutile and anatase crystals in a Cartesian coordinate system. The optimization process was carried out on all the structures using the PM7 semi-empirical calculation. The calculations carried out by the ab-initio methods used the coordinates optimized by the semi-empirical method. In each structure, all the atoms and their respective coordinates (x, y, z) were optimized, looking for regions with the lowest energy level.

5. RESULTS AND DISCUSSION

Based on the methodology proposed in this study, nanotube models were generated. Some of these can be seen in Figures 4.1 to 4.4 (rutile nanotubes) and Figures 4.5 to 4.8 (anatase nanotubes).

Figure 4.1 - Side view of rutile nanotube plane with 1 level generated

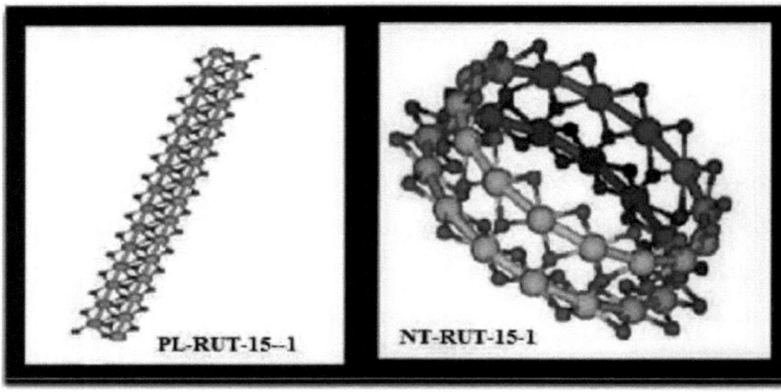

Figure 4.2 - Side view of rutile nanotube plane with 2 levels generated

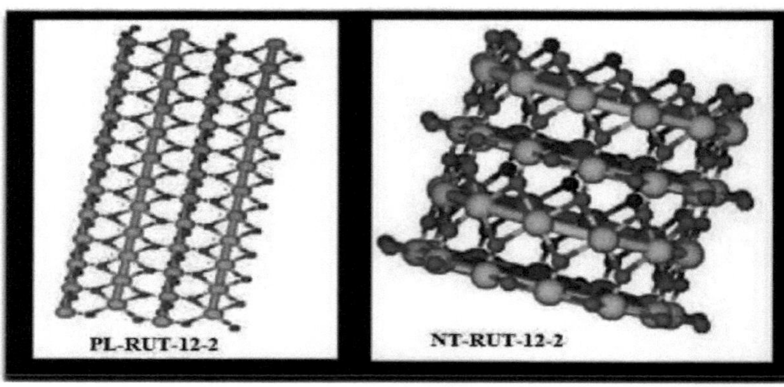

Figure 4.3 - Side view of rutile nanotube plane with 3 levels generated

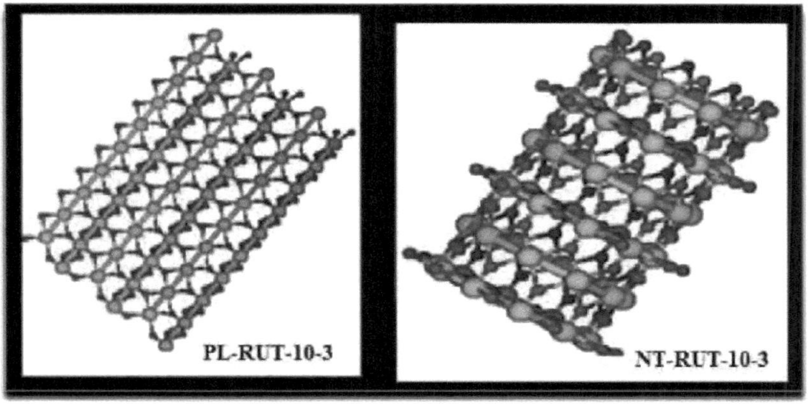

Figure 4.4 - Top view of rutile nanotubes

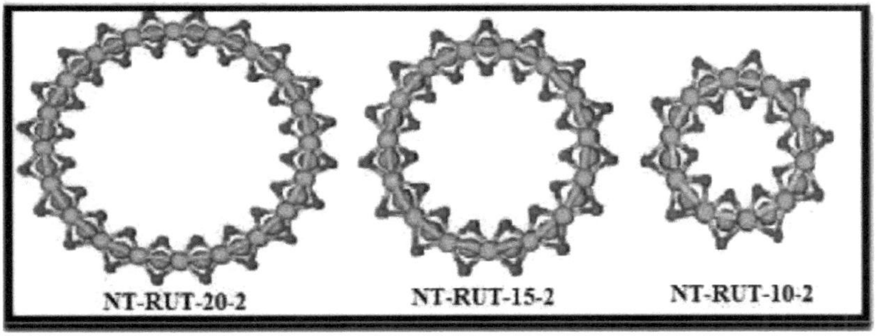

Figure 4.5 - Side view of anatase plane and nanotube with 1 level generated

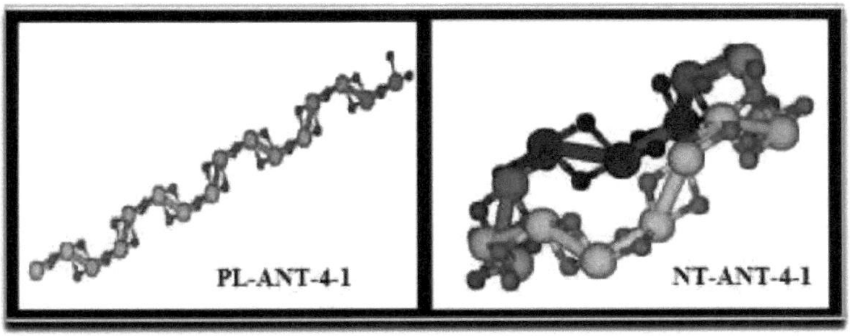

Figure 4.6 - Side view of anatase plane and nanotube with 2 levels generated

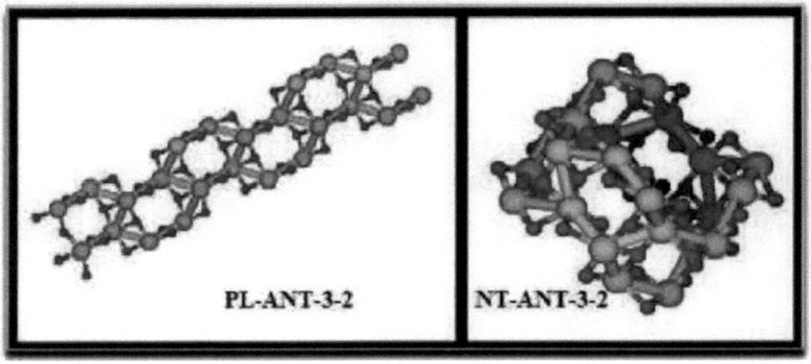

Figure 4.7 - Side view of anatase plane and nanotube with 3 levels generated

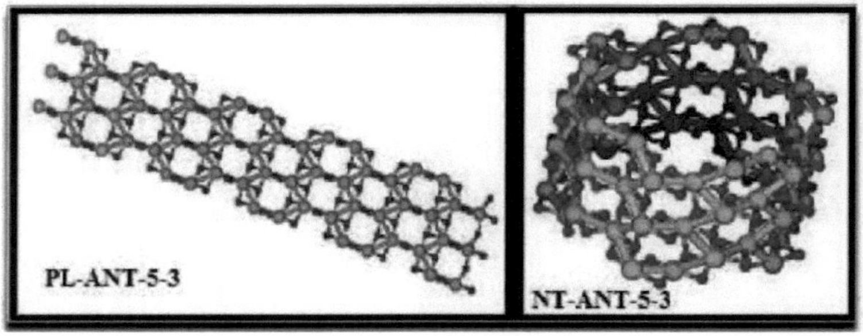

Figure 4.8 - Top view of anatase nanotubes

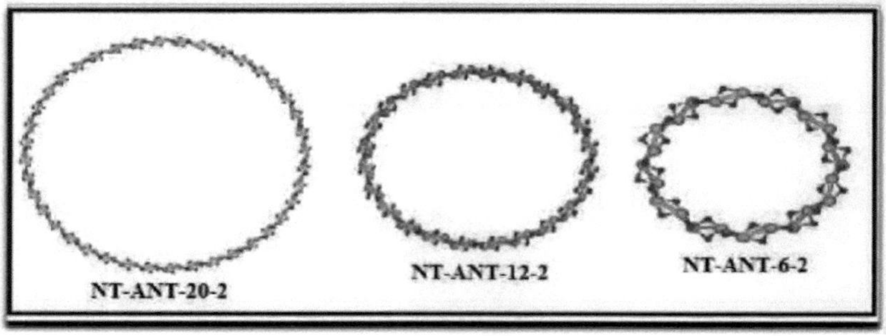

Figures 4.9 and 4.10 illustrate the result of the optimization process that was carried out in the proposed methodology, in which the same structure is illustrated, one with optimization and the other without optimization. The results of the Total Energy of the structures (Figure 4.9) show that the optimized structure has a lower energy level, i.e. greater stability compared to the structure with the initial geometry.

Figure 4.9 - Side view of the NT-ANT-10-3 and NT-ANT-10-3-ot naotubes in the PM7 semi-empirical method.

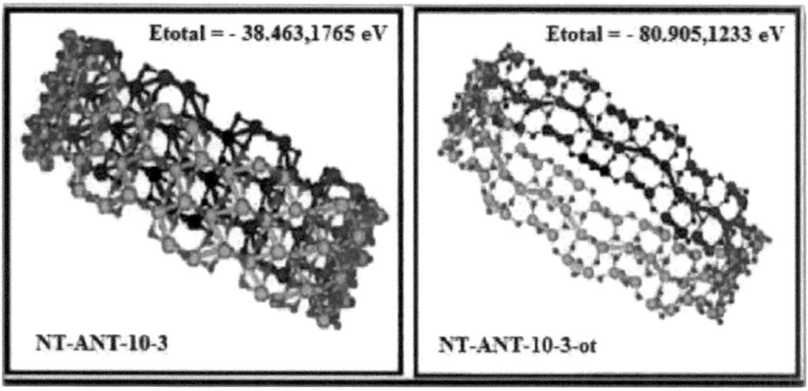

Figure 4.10 - Top view of the nanotubes, NT-ANT-10-3 and NT-ANT-10-3-ot

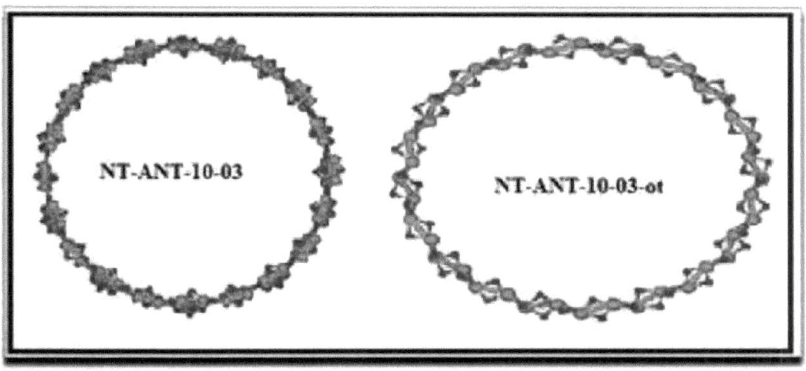

5.1. ENERGY VARIATION ANALYSIS

The energy variation of the structures was analyzed using models optimized by the semi-empirical method (PM7), using the MOPAC software version 2012. Calculations from *ab-initio* methods (HF and DFT) were also used, using the Cartesian coordinates optimized by the semi-empirical method, and were run in the GAUSSIAN software version 2003, in the 6-31G and 6-311G bases.

136 structures (planes and nanotubes) of rutile and anatase were obtained in a semi-empirical method. Some nanotubes, after optimization on a PM7 basis and 12

structures (anatase nanotubes) on an *ab-initio* method. Table 4.1 shows all the calculated structures.

Table 4.1 - Table of structures that were calculated and optimized using the semi-empirical method (PM7)

ANATASE							
	Structure name	**Molecular Formula**	**(Ti02)4 n**	**(Ti02)4 m**	**(Ti02)4 Total**	**(TiO2) Total**	**Number of atoms**
1	NT-ANT-2-1-ot	{[(TiO2)4]2}1	2	1	2	8	24
2	NT-ANT-3-1-ot	{[(TiO2)4]3}1	3	1	3	12	36
3	NT-ANT-4-1-ot	{[(TiO2)4]4}1	4	1	4	16	48
4	NT-ANT-5-1-ot	{[(TiO2)4]5}1	5	1	5	20	60
5	NT-ANT-6-1-ot	{[(TiO2)4]6}1	6	1	6	24	72
6	NT-ANT-7-1-ot	{[(TiO2)4]7}1	7	1	7	28	84
7	NT-ANT-8-1-ot	{[(TiO2)4]8}1	8	1	8	32	96
8	NT-ANT-9-1-ot	{[(TiO2)4]9}1	9	1	9	36	108
9	NT-ANT-11-1-ot	{[(TiO2)4]11}1	11	1	11	44	132
10	NT-ANT-12-1-ot	{[(TiO2)4]12}1	12	1	12	48	144
11	NT-ANT-13-1-ot	{[(TiO2)4]13}1	13	1	13	52	156
12	NT-ANT-14-1-ot	{[(TiO2)4]14}1	14	1	14	56	168
13	NT-ANT-15-1-ot	{[(TiO2)4]15}1	15	1	15	60	180
14	NT-ANT-16-1-ot	{[(TiO2)4]16}1	16	1	16	64	192
15	NT-ANT-17-1-ot	{[(TiO2)4]17}1	17	1	17	68	204
16	NT-ANT-18-1-ot	{[(TiO2)4]18}1	18	1	18	72	216
17	NT-ANT-19-1-ot	{[(TiO2)4]19}1	19	1	19	76	228
18	NT-ANT-20-1-ot	{[(TiO2)4]20}1	20	1	20	80	240
19	PL-ANT-2-1-ot	{[(TiO2)4]2}1	2	1	2	8	24
20	PL-ANT-3-1-ot	{[(TiO2)4]3}1	3	1	3	12	36
21	PL-ANT-4-1-ot	{[(TiO2)4]4}1	4	1	4	16	48
22	PL-ANT-5-1-ot	{[(TiO2)4]5}1	5	1	5	20	60
23	PL-ANT-6-1-ot	{[(TiO2)4]6}1	6	1	6	24	72
24	PL-ANT-7-1-ot	{[(TiO2)4]7}1	7	1	7	28	84
25	PL-ANT-8-1-ot	{[(TiO2)4]8}1	8	1	8	32	96
26	PL-ANT-9-1-ot	{[(TiO2)4]9}1	9	1	9	36	108
27	PL-ANT-10-1-ot	{[(TiO2)4]10}1	10	1	10	40	120
28	PL-ANT-11-1-ot	{[(TiO2)4]11}1	11	1	11	44	132
29	PL-ANT-12-1-ot	{[(TiO2)4]12}1	12	1	12	48	144

30	PL-ANT-13-1-ot	{[(TiO2)4]13}1	13	1	13	52	156
31	PL-ANT-14-1-ot	{[(TiO2)4]14}1	14	1	14	56	168
32	PL-ANT-15-1-ot	{[(TiO2)4]15}1	15	1	15	60	180
33	PL-ANT-16-1-ot	{[(TiO2)4]16}1	16	1	16	64	192
34	PL-ANT-17-1-ot	{[(TiO2)4]17}1	17	1	17	68	204
35	PL-ANT-18-1-ot	{[(TiO2)4]18}1	18	1	18	72	216
36	PL-ANT-19-1-ot	{[(TiO2)4]19}1	19	1	19	76	228
37	PL-ANT-20-1-ot	{[(TiO2)4]20}1	20	1	20	80	240
38	NT-ANT-2-2-ot	{[(TiO2)4]2}2	2	2	4	16	48
39	NT-ANT-3-2-ot	{[(TiO2)4]3}2	3	2	6	24	72
40	NT-ANT-4-2-ot	{[(TiO2)4]4}2	4	2	8	32	96
41	NT-ANT-5-2-ot	{[(TiO2)4]5}2	5	2	10	40	120
42	NT-ANT-6-2-ot	{[(TiO2)4]6}2	6	2	12	48	144
43	NT-ANT-7-2-ot	{[(TiO2)4]7}2	7	2	14	56	168
44	NT-ANT-8-2-ot	{[(TiO2)4]8}2	8	2	16	64	192
45	NT-ANT-9-2-ot	{[(TiO2)4]9}2	9	2	18	72	216
46	NT-ANT-10-2-ot	{[(TiO2)4]10}2	10	2	20	80	240
47	NT-ANT-11-2-ot	{[(TiO2)4]11}2	11	2	22	88	264
48	NT-ANT-12-2-ot	{[(TiO2)4]12}2	12	2	24	96	288
49	NT-ANT-13-2-ot	{[(TiO2)4]13}2	13	2	26	104	312
50	NT-ANT-14-2-ot	{[(TiO2)4]14}2	14	2	28	112	336
51	NT-ANT-15-2-ot	{[(TiO2)4]15}2	15	2	30	120	360
52	PL-ANT-2-2-ot	{[(TiO2)4]2}2	2	2	4	16	48
53	PL-ANT-3-2-ot	{[(TiO2)4]3}2	3	2	6	24	72
54	PL-ANT-4-2-ot	{[(TiO2)4]4}2	4	2	8	32	96
55	PL-ANT-5-2-ot	{[(TiO2)4]5}2	5	2	10	40	120
56	PL-ANT-6-2-ot	{[(TiO2)4]6}2	6	2	12	48	144
57	PL-ANT-7-2-ot	{[(TiO2)4]7}2	7	2	14	56	168
58	PL-ANT-8-2-ot	{[(TiO2)4]8}2	8	2	16	64	192
59	PL-ANT-9-2-ot	{[(TiO2)4]9}2	9	2	18	72	216
60	PL-ANT-10-2-ot	{[(TiO2)4]10}2	10	2	20	80	240
61	PL-ANT-11-2-ot	{[(TiO2)4]11}2	11	2	22	88	264
62	PL-ANT-12-2-ot	{[(TiO2)4]12}2	12	2	24	96	288
63	PL-ANT-13-2-ot	{[(TiO2)4]13}2	13	2	26	104	312
64	PL-ANT-14-2-ot	{[(TiO2)4]14}2	14	2	28	112	336
65	NT-ANT-2-3-ot	{[(TiO2)4]2}3	2	3	6	24	72
66	NT-ANT-3-3-ot	{[(TiO2)4]3}3	3	3	9	36	108
67	NT-ANT-4-3-ot	{[(TiO2)4]4}3	4	3	12	48	144

68	NT-ANT-5-3-ot	{[(TiO2)4]5}3	5	3	15	60	180
69	NT-ANT-6-3-ot	{[(TiO2)4]6}3	6	3	18	72	216
70	NT-ANT-7-3-ot	{[(TiO2)4]7}3	7	3	21	84	252
71	NT-ANT-8-3-ot	{[(TiO2)4]8}3	8	3	24	96	288
72	NT-ANT-9-3-ot	{[(TiO2)4]9}3	9	3	27	108	324
73	NT-ANT-10-3-ot	{[(TiO2)4]10}3	10	3	30	120	360
74	NT-ANT-11-3-ot	{[(TiO2)4]11}3	11	3	33	132	396
75	NT-ANT-12-3-ot	{[(TiO2)4]12}3	12	3	36	144	432
76	NT-ANT-13-3-ot	{[(TiO2)4]13}3	13	3	39	156	468
77	NT-ANT-17-3-ot	{[(TiO2)4]17}3	17	3	51	204	612
78	NT-ANT-19-3-ot	{[(TiO2)4]19}3	19	3	57	228	684
79	NT-ANT-20-3-ot	{[(TiO2)4]20}3	20	3	60	240	720
80	PL-ANT-2-3-ot	{[(TiO2)4]2}3	2	3	6	24	72
81	PL-ANT-3-3-ot	{[(TiO2)4]3}3	3	3	9	36	108
82	PL-ANT-4-3-ot	{[(TiO2)4]4}3	4	3	12	48	144
83	PL-ANT-5-3-ot	{[(TiO2)4]5}3	5	3	15	60	180
84	PL-ANT-7-3-ot	{[(TiO2)4]7}3	7	3	21	84	252
85	PL-ANT-13-3-ot	{[(TiO2)4]13}3	13	3	39	156	468
RUTILO							
	Structure name	**Molecular Formula**	**(TiO2)2 (YZ)**	**(TiO2)2 (X)**	**(TiO2)2 Total**	**(TiO2) Total**	**Number of atoms**
86	NT-RUT-6-1-ot	{[(TiO2)2]6}1	6	1	6	12	36
87	NT-RUT-7-1-ot	{[(TiO2)2]7}1	7	1	7	14	42
88	NT-RUT-8-1-ot	{[(TiO2)2]8}1	8	1	8	16	48
89	NT-RUT-9-1-ot	{[(TiO2)2]9}1	9	1	9	18	54
90	NT-RUT-10-1-ot	{[(TiO2)2]10}1	10	1	10	20	60
91	NT-RUT-11-1-ot	{[(TiO2)2]11}1	11	1	11	22	66
92	NT-RUT-12-1-ot	{[(TiO2)2]12}1	12	1	12	24	72
93	NT-RUT-13-1-ot	{[(TiO2)2]13}1	13	1	13	26	78
94	NT-RUT-14-1-ot	{[(TiO2)2]14}1	14	1	14	28	84
95	NT-RUT-15-1-ot	{[(TiO2)2]15}1	15	1	15	30	90
96	NT-RUT-16-1-ot	{[(TiO2)2]16}1	16	1	16	32	96
97	NT-RUT-17-1-ot	{[(TiO2)2]17}1	17	1	17	34	102
98	NT-RUT-18-1-ot	{[(TiO2)2]18}1	18	1	18	36	108
99	NT-RUT-19-1-ot	{[(TiO2)2]19}1	19	1	19	38	114
100	NT-RUT-20-1-ot	{[(TiO2)2]20}1	20	1	20	40	120
101	PL-RUT-6-1-ot	{[(TiO2)2]6}1	6	1	6	12	36
102	PL-RUT-7-1-ot	{[(TiO2)2]7}1	7	1	7	14	42

103	PL-RUT-8-1-ot	{[(TiO2)2]8}1	8	1	8	16	48
104	PL-RUT-9-1-ot	{[(TiO2)2]9}1	9	1	9	18	54
105	PL-RUT-10-1-ot	{[(TiO2)2]10}1	10	1	10	20	60
106	PL-RUT-11-1-ot	{[(TiO2)2]11}1	11	1	11	22	66
107	PL-RUT-12-1-ot	{[(TiO2)2]12}1	12	1	12	24	72
108	PL-RUT-13-1-ot	{[(TiO2)2]13}1	13	1	13	26	78
109	PL-RUT-14-1-ot	{[(TiO2)2]14}1	14	1	14	28	84
110	PL-RUT-16-1-ot	{[(TiO2)2]16}1	16	1	16	32	96
111	PL-RUT-17-1-ot	{[(TiO2)2]17}1	17	1	17	34	102
112	PL-RUT-19-1-ot	{[(TiO2)2]19}1	19	1	19	38	114
113	NT-RUT-6-2-ot	{[(TiO2)2]6}2	6	2	12	24	72
114	NT-RUT-9-2-ot	{[(TiO2)2]9}2	9	2	18	36	108
115	NT-RUT-10-2-ot	{[(TiO2)2]10}2	10	2	20	40	120
116	NT-RUT-11-2-ot	{[(TiO2)2]11}2	11	2	22	44	132
117	NT-RUT-12-2-ot	{[(TiO2)2]12}2	12	2	24	48	144
118	NT-RUT-13-2-ot	{[(TiO2)2]13}2	13	2	26	52	156
119	NT-RUT-16-2-ot	{[(TiO2)2]16}2	16	2	32	64	192
120	NT-RUT-17-2-ot	{[(TiO2)2]17}2	17	2	34	68	204
121	NT-RUT-20-2-ot	{[(TiO2)2]20}2	20	2	40	80	240
122	PL-RUT-6-2-ot	{[(TiO2)2]6}2	6	2	12	24	72
123	PL-RUT-7-2-ot	{[(TiO2)2]7}2	7	2	14	28	84
124	PL-RUT-8-2-ot	{[(TiO2)2]8}2	8	2	16	32	96
125	PL-RUT-9-2-ot	{[(TiO2)2]9}2	9	2	18	36	108
126	PL-RUT-11-2-ot	{[(TiO2)2]11}2	11	2	22	44	132
127	NT-RUT-6-3-ot	{[(TiO2)2]6}3	6	3	18	36	108
128	NT-RUT-8-3-ot	{[(TiO2)2]8}3	8	3	24	48	144
129	NT-RUT-9-3-ot	{[(TiO2)2]9}3	9	3	27	54	162
130	NT-RUT-10-3-ot	{[(TiO2)2]10}3	10	3	30	60	180
131	NT-RUT-11-3-ot	{[(TiO2)2]11}3	11	3	33	66	198
132	NT-RUT-12-3-ot	{[(TiO2)2]12}3	12	3	36	72	216
133	NT-RUT-13-3-ot	{[(TiO2)2]13}3	13	3	39	78	234
134	PL-RUT-6-3-ot	{[(TiO2)2]6}3	6	3	18	36	108
135	PL-RUT-7-3-ot	{[(TiO2)2]7}3	7	3	21	42	126
136	PL-RUT-10-3-ot	{[(TiO2)2]10}3	10	3	30	60	180

In order to compare the energy values obtained from nanotubes and planes, it is necessary to evaluate the variation in energy between the models, as the analysis of the total energy values cannot be used for comparative purposes. First of all, the energy of a unit of (TiO_2) was taken as a reference, which was calculated to be -669.78524 eV. The variation was obtained from the chemical equation (4.1):

$$(nxm)[(TiO_2)Y] \rightleftarrows \{[(TiO_2)_Y]_n\}_m \quad (4.1)$$

Considering that the variation is the difference between the energy of the final product (formed structure) and the energy of the initial structure (TiO_2) multiplied by n and m units of (TiO_2) that have been added to the structure, we have Equation 4.2, which tells us the energy variation of a model.

ΔE = E(Final Structure y with n and m levels) - (yxnx m)E(TiO2 Unit)

$$\Delta E = E\{[(TiO_2)_Y]_n\}_m - [(mxn) \times Y] \times E(TiO_2) \quad (4.2)$$

Where the variable Y is the index that identifies the structure (rutile, anatase), with Y=2 for rutile and Y=4 for anatase.

5.1.1. Analysis of energy variation in anatase nanotubes at 3 levels (m=1,2 and 3).

47 models of 1, 2 and 3-level anatase nanotubes were obtained, with 18 models of 1-level anatase nanotubes showing the following results in Table 4.2 and the graph in Figure 4.11:

Table 4.2 - Table of variation of energies found in 1-level anatase nanotubes in semi-empirical method (PM7)

Molecular Formula	**n(TiO_2)4 (YZ)**	**m(TiO_2)4 (X)**	**(TiO_2) Total**	**Total Energy (eV)**	**ΔE (eV)**	**ΔE/(TiO_2) Total (eV)**
{[(TiO2)4]2}1	2	1	8	-5384,8404	-26,5585	-3,3198
{[(TiO2)4]3}1	3	1	12	-8071,8975	-34,4746	-2,8729
{[(TiO2)4]4}1	4	1	16	-10763,3135	-46,7496	-2,9219
{[(TiO2)4]5}1	5	1	20	-13454,4691	-58,7643	-2,9382
{[(TiO2)4]6}1	6	1	24	-16145,5155	-70,6697	-2,9446
{[(TiO2)4]7}1	7	1	28	-18836,5180	-82,5312	-2,9475
{[(TiO2)4]8}1	8	1	32	-21527,7448	-94,6171	-2,9568

{[(TiO2)4]9}1	9	1	36	-24218,7040	-106,4350	-2,9565
{[(TiO2)4]11}1	11	1	44	-29600,6832	-130,1330	-2,9576
{[(TiO2)4]12}1	12	1	48	-32291,6598	-141,9680	-2,9577
{[(TiO2)4]13}1	13	1	52	-34982,6786	-153,8460	-2,9586
{[(TiO2)4]14}1	14	1	56	-37673,6939	-165,7200	-2,9593
{[(TiO2)4]15}1	15	1	60	-40364,7064	-177,5920	-2,9599
{[(TiO2)4]16}1	16	1	64	-43055,7168	-189,4610	-2,9603
{[(TiO2)4]17}1	17	1	68	-45746,7254	-201,3290	-2,9607
{[(TiO2)4]18}1	18	1	72	-48437,7325	-213,1950	-2,9610
{[(TiO2)4]19}1	19	1	76	-51128,7384	-225,0600	-2,9613
{[(TiO2)4]20}1	20	1	80	-53819,7413	-236,9220	-2,9615

Figure 4.11 - Graph of the variation of energy in eV by the number of units n [$(TiO_2)_4$] of the anatase nanotubes of the 1^0 level (m=1) by the semi-empirical method (PM7).

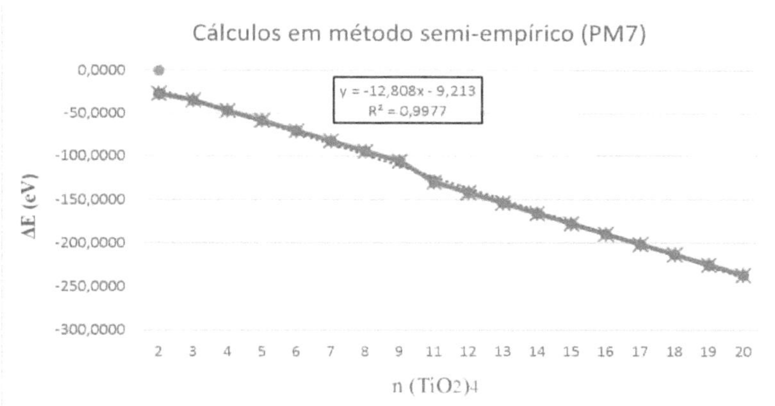

Based on the **AE** values and the number of levels n shown in Table 4.2 and the graph in Figure 4.11, it can be seen that there is a decrease in the value of the energy variation as the number of (TiO)$_{24}$ units increases in the YZ plane. This implies that for anatase nanotubes formed with 1 level (m=1), it was found that increasing the diameter of the nanotube led to the formation of more stable structures. One indication of this behavior is the greater tension in the smaller radius nanotubes, which makes the smaller diameter structures more unstable than the larger ones.

The trend in the energy variation of anatase nanotubes from 2 to 20 units of $(TiO_2)_4$ in the YZ plane showed a good linear behavior, which can be seen from the trend line that is practically coincident with the **AE** slope line and the correlation factor (R^2 = 0.9977), but it cannot be said that linear regression can describe the behavior of nanotubes from 20 units onwards.

The results show that, among the models obtained, the most stable structure is the one with the largest diameter, which is the NT-ANT-20-1-ot cluster. However, given the trend observed, it is likely that there are clusters with larger diameters of more stable nanotubes.

The 13 anatase nanotube models with 2 levels obtained had the following values, described in Table 4.3 and in the trend curve graph in Figure 4.12:

Table 4.3 - Range of energies found in 2-level anatase nanotubes in semi-empirical method (PM7)

Molecular Formula	**n(Ti02)4 (YZ)**	**m(TiO2)4 (X)**	**(Ti02) (Total)**	**Total Energy (eV)**	**AE (eV)**	**AE/(TiO2) Total (eV)**
{[(TiO2)4]2}2	2	2	16	-10790,5176	-73,9538	-4,6221
{[(TiO2)4]3}2	3	2	24	-16184,8146	-109,9690	-4,5820
{[(TiO2)4]4}2	4	2	32	-21579,0746	-145,9470	-4,5608
{[(TiO2)4]5}2	5	2	40	-26965,6826	-174,2730	-4,3568
{[(TiO2)4]6}2	6	2	48	-32359,2663	-209,5750	-4,3661
{[(TiO2)4]7}2	7	2	56	-37752,6295	-244,6560	-4,3689
{[(TiO2)4]8}2	8	2	64	-43145,9670	-279,7120	-4,3705
{[(TiO2)4]9}2	9	2	72	-48557,7326	-333,1950	-4,6277
{[(TiO2)4]10}2	10	2	80	-53949,8782	-367,0590	-4,5882
{[(TiO2)4]11}2	11	2	88	-59345,6173	-404,5160	-4,5968
{[(TiO2)4]12}2	12	2	96	-64740,7023	-441,3190	-4,5971
{[(TiO2)4]13}2	13	2	104	-70135,7536	-478,0890	-4,5970
{[(TiO2)4]14}2	14	2	112	-75530,7654	-514,8180	-4,5966

Figure 4.12 - Graph of the variation of energy in eV by the number of units n [(TiO2)4] of the anatase nanotubes of the 2º level (m=2) by the semi-empirical method (PM7).

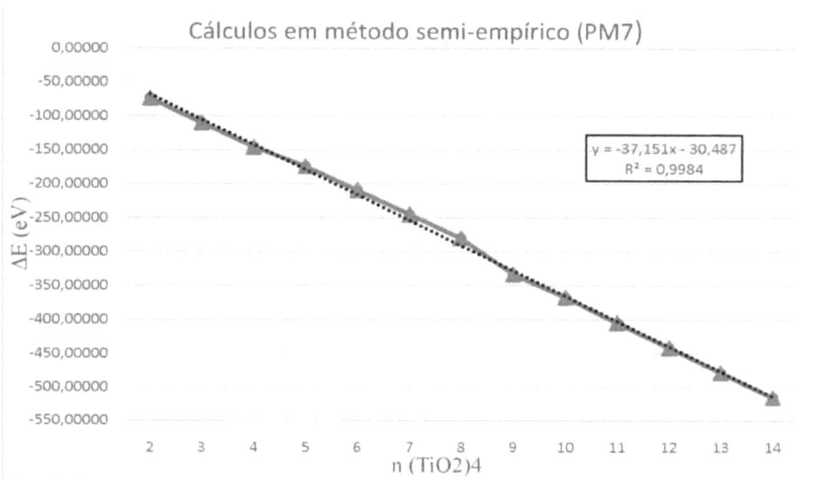

From the values of **AE** and the number of levels n shown in Table 4.3 and the graph in Figure 4.2, it can be seen that there is a decrease in the value of the energy variation as the number of $(TiO_2)_4$ units in the YZ plane increases. This behavior is similar to that of the 1-level nanotubes, showing that the increase in nanotube diameter led to the formation of more stable structures. The growth of the nanotube with 2 levels present did not imply a difference in the **AE** trend.

The trend in the energy variation of nanotubes from 2 to 14 $(TiO_2)_4$ units in the YZ plane also showed good linear behavior, due to the trend line being practically coincident with **the AE** slope line and the correlation factor (R^2 = 0.9984) being very close to 1.000. Nor can it be concluded that the linear regression can describe the behavior of the 2-level nanotubes from 14 units onwards, as in the first level.

Among the structures calculated, no stabilization region was observed. The structure with the largest diameter (NT-ANT-14-2-ot) appeared to be the most stable structure, but the stabilization trend suggests that there may be more stable structures.

With the 3^0 level structures, 15 nanotubes were obtained, which showed the following values described in Table 4.4 and in the trend curve graph in Figure 4.13:

Table 4.4 - Table of variation of energies found in 3-level anatase nanotubes in semi-empirical method (PM7)

Molecular Formula	n(Ti02)4 (YZ)	m(TiO2)4 (X)	(TiO2) Total	Total Energy (eV)	AE (eV)	AE/(TiO2) Total (eV)
{[(TiO2)4]2}3	2	3	24	-16196,9562	-122,1100	-5,0879
{[(TiO2)4]3}3	3	3	36	-24295,2048	-182,9360	-5,0816
{[(TiO2)4]4}3	4	3	48	-32381,3576	-231,6660	-4,8264
{[(TiO2)4]5}3	5	3	60	-40480,6684	-293,5540	-4,8926
{[(TiO2)4]6}3	6	3	72	-48547,6770	-323,1400	-4,4881
{[(TiO2)4]7}3	7	3	84	-56634,6294	-372,6690	-4,4365
{[(TiO2)4]8}3	8	3	96	-64724,8310	-425,4480	-4,4318
{[(TiO2)4]9}3	9	3	108	-72814,9914	-478,1850	-4,4276
{[(TiO2)4]10}3	10	3	120	-80905,1233	-530,8940	-4,4241
{[(TiO2)4]11}3	11	3	132	-89009,3930	-597,7410	-4,5283
{[(TiO2)4]12}3	12	3	144	-97085,3311	-636,2570	-4,4185
{[(TiO2)4]13}3	13	3	156	-105184,3641	-697,8670	-4,4735
{[(TiO2)4]17}3	17	3	204	-137547,0305	-910,8420	-4,4649
{[(TiO2)4]19}3	19	3	228	-153696,9783	-985,9440	-4,3243
{[(TiO2)4]20}3	20	3	240	-161785,9274	-1.037,4700	-4,3228

Figure 4.13 - Graph of the variation of energy in eV by the number of units n [(TiO2)4] of the anatase nanotubes of the 3º level (m=3) by the semi-empirical method (PM7).

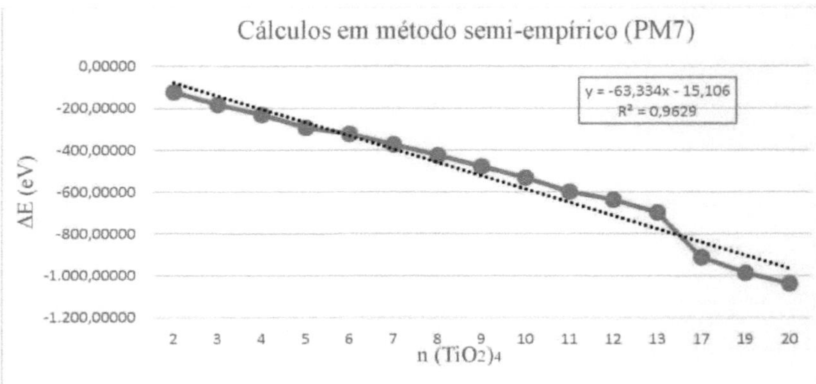

Anatase nanotubes with 3 levels behaved similarly to nanotubes with 1 and 2 levels. From the values of **AE** and the number of levels n shown in Table 4.4 and the graph in Figure 4.3, it can be seen that there is also a decrease in the value of the energy variation as the diameter increases, forming more stable structures. The presence of 3 levels, increasing the length of the nanotube, as was seen in the 2-level nanotubes, did not influence the change in behavior, i.e. the smaller diameter structures remained more unstable than the larger diameter ones.

The trend in the energy variation of the nanotubes at this level has already shown some differences from the other 2 previous levels. It showed a linear behavior, but the linear trend line is not coincident with the **ΔE** slope line and the correlation factor (R^2 = 0.9629) is more distant from 1.000 compared to the nanotubes of the I^0 and 2^0 levels. The data show that the NT-ANT-20-3-ot structure is the most stable of the 30-level nanotubes. Similarly, linear regression cannot describe the behavior of 3-level nanotubes from 20 units. The 3 nanotube stability trend curves are shown in Figure 4.14.

Figure 4.14 - Graph of the energy variation in eV by the number of n [(TiO_2)4] units of the anatase nanotubes at the 3 levels using the semi-empirical method (PM7)

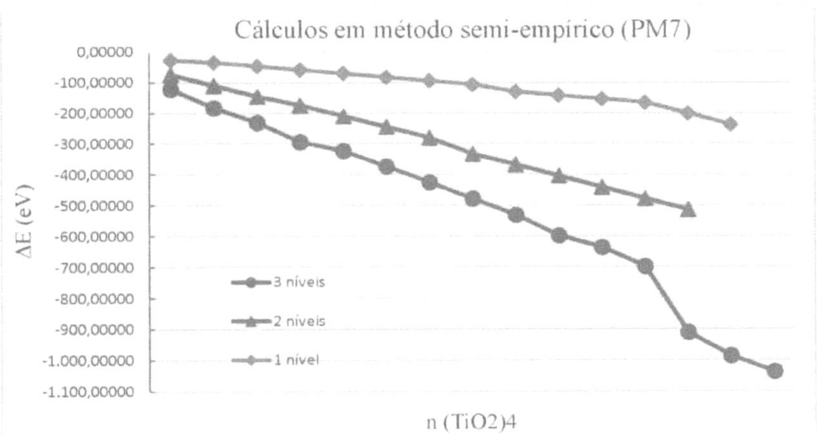

The graph in Figure 4.14 outlines the energy variation curves of the different structures in the 3 levels of anatase nanotubes. All 3 curves show similar behavior, as described above. However, comparing the structures of the 3 levels using the unit reference (TiO_2) is only possible for structures with the same number of levels (evaluation of the energy variation as a function of diameter and curvature) or for structures with the same number of TiO_2 units, i.e. atoms.

5.1.2. Analysis of energy variation in rutile nanotubes for m=1,2 and 3.

We obtained 31 models of rutile nanotubes, with 1, 2 and 3 levels, and 15 models of anatase nanotubes with 1 level, which showed the following results described in Table 4.5:

Table 4.5 - Table of variation of energies found in 1-level rutile nanotubes in semi-empirical method (PM7)

Molecular Formula	n(TiO_2)2 (YZ)	m(TiO_2)2 (X)	(TiO_2) (Total)	Total Energy (eV)	AE (eV)	AE/(TiO_2) Total (eV)
{[(TiO2)2]6}1	6	1	12	-8083,9625	-46,5396	-3,8783
{[(TiO2)2]7}1	7	1	14	-9439,3329	-62,3395	-4,4528
{[(TiO2)2]8}1	8	1	16	-10775,4015	-58,8377	-3,6774
{[(TiO2)2]9}1	9	1	18	-12123,0293	-66,8950	-3,7164
{[(TiO2)2]10}1	10	1	20	-13482,6420	-86,9372	-4,3469
{[(TiO2)2]11}1	11	1	22	-14824,2167	-88,9414	-4,0428
{[(TiO2)2]12}1	12	1	24	-16176,3601	-101,5140	-4,2298
{[(TiO2)2]13}1	13	1	26	-17528,5064	-114,0900	-4,3881
{[(TiO2)2]14}1	14	1	28	-18866,0994	-112,1130	-4,0040
{[(TiO2)2]15}1	15	1	30	-20221,5736	-128,0160	-4,2672
{[(TiO2)2]16}1	16	1	32	-21566,2821	-133,1540	-4,1611
{[(TiO2)2]17}1	17	1	34	-22915,6795	-142,9810	-4,2053
{[(TiO2)2]18}1	18	1	36	-24257,4000	-145,1310	-4,0314
{[(TiO2)2]19}1	19	1	38	-25614,8040	-162,9650	-4,2886
{[(TiO2)2]20}1	20	1	40	-26971,4363	-180,0270	-4,5007

Figure 4.15 - Graph of the energy variation in eV by the number of units n [(TiO_2)2] of the rutile nanotubes of the 1º level (m=1) by the semi-empirical method (PM7)

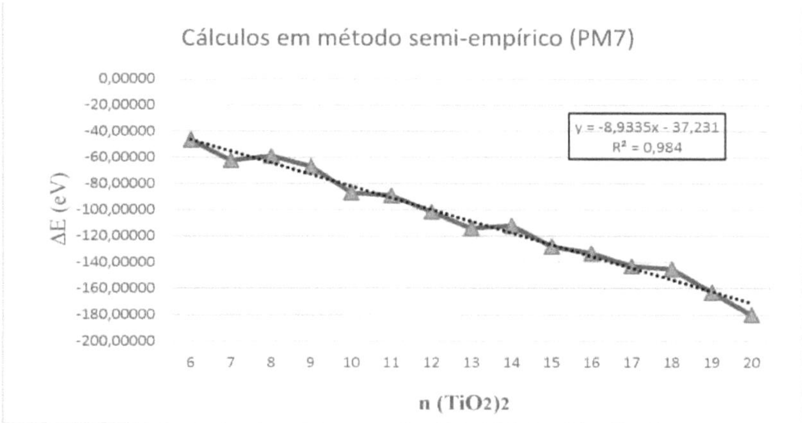

From the **AE** values and the number of n levels shown in Table 4.5, it can be seen that the energy variation decreases as the number of $(TiO_2)_2$ units in the YZ plane increases. This implies that for the rutile nanotubes formed with 1 level (m=1), it was found that the increase in nanotube diameter led to the formation of more stable structures. An indication of this behavior is the greater tension of the smaller radius nanotubes, which makes the smaller diameter structures more unstable than the larger ones. A similar idea to anatase nanotubes. In other words, despite the different structure conformations between rutile and anatase nanotubes, the behavior was similar.

The trend in the energy variation of the nanotubes with the value of n from 6 to 20 units showed a good linear behavior, which can be seen in the trend line in Figure 4.15, which is practically coincident with the slope line of the **AE**, and in the correlation factor (R^2 = 0.984), although it cannot be said that the linear regression can describe the behavior of the nanotubes from 20 units onwards.

According to the results, among the models obtained, the most stable structure is the one with the largest diameter, which is the NT-RUT-20-1-ot cluster. However, given the trend observed, it is likely that there are clusters with larger diameters of more stable nanotubes.

With 02 levels, 09 models of rutile nanotubes were obtained and presented the following values described in Table 4.6:

Table 4.6 - Table of variation of energies found in 2-level rutile nanotubes in semi-empirical method (PM7)

Molecular Formula	n(Ti02)2 (YZ)	m(TiO2)2 (X)	(Ti02) (Total)	Total Energy (eV)	AE (eV)	AE/(TiO2) Total (eV)
{[(TiO2)2]6}2	6	2	24	-16178,4333	-103,5880	-4,3162
{[(TiO2)2]9}2	9	2	36	-24275,7073	-163,4390	-4,5400
{[(TiO2)2]10}2	10	2	40	-26968,2363	-176,8270	-4,4207
{[(TiO2)2]11}2	11	2	44	-29653,6038	-183,0530	-4,1603
{[(TiO2)2]12}2	12	2	48	-32347,5095	-197,8180	-4,1212
{[(TiO2)2]13}2	13	2	52	-35072,8026	-243,9700	-4,6917
{[(TiO2)2]16}2	16	2	64	-43143,0054	-276,7500	-4,3242
{[(TiO2)2]17}2	17	2	68	-45847,1640	-301,7680	-4,4378
{[(TiO2)2]20}2	20	2	80	-53949,5595	-366,7400	-4,5843

From the **AE** values and the number of levels n shown in Table 4.6, it can be seen that the energy variation decreases as the number of $(TiO\ )_{22}$ units in the YZ plane increases. Similarly to the 1-level nanotubes, it was found that increasing the diameter of the nanotube led to the formation of more stable structures. Despite the presence of another level, implying a structure with a greater number of atoms, there was no difference in behavior compared to the 1-level nanotubes. The graph in Figure 4.6 shows a tendency towards stabilization.

Figura 4.16- Graph of the energy variation in eV by the number of units n [(TiO2)2] of the rutile nanotubes of the 2º level (m=2) by the semi-empirical method (PM7)

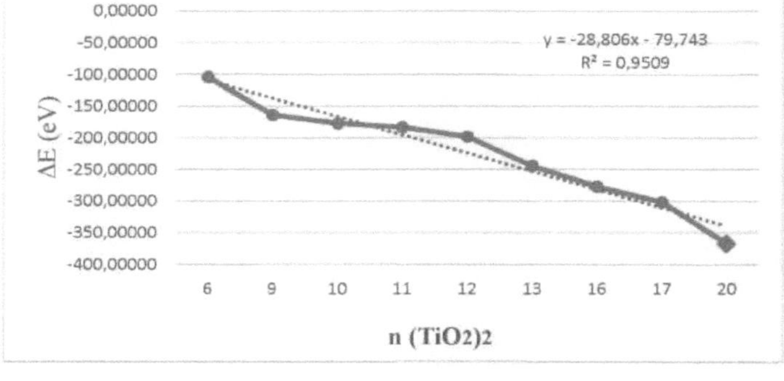

The trend in the energy variation of the nanotubes with the value of n from 2 to 14 units also showed linear behavior, but the behavior is not so linear when compared to the trend curves of the anatase and rutile nanotubes of the I^0 level, this being observed by the correlation factor ($R^2 = 0.9509$).

Linear regression no longer describes the behavior of 2-level nanotubes as accurately as it does at 20 units.

The behavior of the curve probably indicates the existence of more stable structures than NT-RUT-20-2-ot1.

With the 30-level nanotubes, 7 nanotubes were obtained, which showed the values described in Table 4.7.

Table 4.7 - Table of variation of energies found in 3-level rutile nanotubes in semi-empirical method (PM7)

Molecular Formula	**n(TiO2)2 (YZ)**	**m(TiO2)2 (X)**	**(TiO2) (Total)**	**Total Energy (eV)**	**AE (eV)**	**AE/(TiO2) Total (eV)**
{[(TiO2)2]6}3	6	3	36	-24277,3811	-165,1120	-4,5864
{[(TiO2)2]8}3	8	3	48	-32354,8632	-205,1720	-4,2744
{[(TiO2)2]9}3	9	3	54	-36422,3033	-253,9000	-4,7019
{[(TiO2)2]10}3	10	3	60	-40460,4892	-273,3750	-4,5563
{[(TiO2)2]11}3	11	3	66	-44491,3285	-285,5030	-4,3258
{[(TiO2)2]12}3	12	3	72	-48551,1762	-326,6390	-4,5367
{[(TiO2)2]13}3	13	3	78	-52593,1000	-349,8510	-4,4853

Figura 4.17- Graph of the energy variation in eV by the number of units n [(TiO2)2] of the rutile nanotubes of the 3º level (m=3) by the semi-empirical method (PM7)

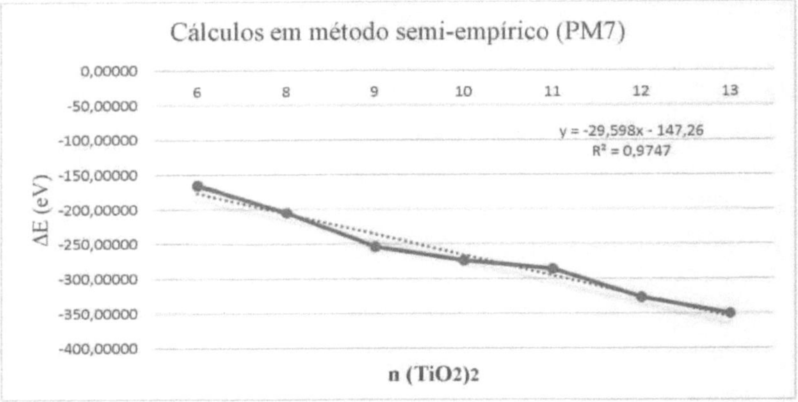

The rutile nanotubes with 3 levels showed similar behavior to the nanotubes with 1 and 2 levels. From the values of **AE** and the number of levels n shown in table 4.7 and the graph in figure 4.7, it can be seen that there is also a decrease in the value of the energy variation as the diameter increases, forming more stable structures. The presence of 3 levels, increasing the length of the nanotube, did not alter the fact that there was greater tension between the atoms of the smaller radius nanotubes. The smaller diameter structures continued to be more unstable than the larger ones.

The trend in the energy variation of the nanotubes at this level showed a linear behavior, similar to the 2^0 level with correlation factor ($R^2 = 0.9629$). The data show that the NT-RUT-20-3- ot structure is the most stable. It cannot be said that linear regression can describe the behavior of 3-level nanotubes starting at 13 units. The graph in figure 4.18 shows the behavior of the 3-level stability trends.

Figure 4.18 - Graph of the energy variation in eV by the number of n [$(TiO_2)_2$] units of the anatase nanotubes at the 3 levels using the semi-empirical method (PM7)

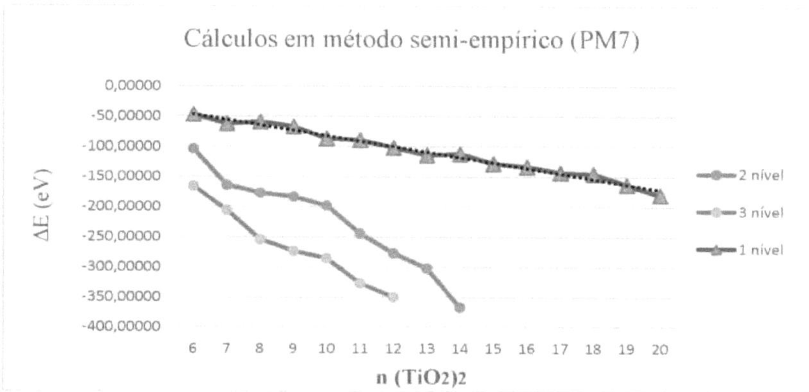

The graph in Figure 4.18 outlines the energy variation curves of the different structures in the 3 levels of rutile nanotubes. All 3 curves show similar behavior, as described above. In the same way as described in the analysis of anatase nanotubes, comparisons between the structures of the 3 levels using the unit reference (TiO_2) are only possible for structures with the same number of levels (evaluation of energy variation as a function of diameter and curvature) or between structures with the same number of TiO_2 units, i.e. atoms.

5.1.3. Comparison between anatase and rutile nanotube models

By comparing anatase and rutile nanotube values as a function of TiO units$_2$, this comparison becomes valid for models with the same number of atoms. Table 4.8 shows the selected models:

Table 4.8 - Table of energy variations between rutile and anatase nanotubes with the same number of units in the semi-empirical method (PM7)

Nanotube	Molecular Formula	n(Ti02)	m(TiO2)	(Ti02) Total	N⁰ atoms	AE (eV)
NT-ANT-12-1-ot-1	{[(TiO2)4]12}1	12	1	48	144	-133,1540
NT-RUT-12-2-ot	{[(TiO2)2]12}2	12	2	48	144	-94,6171
NT-ANT-8-1-ot	{[(TiO2)4]8}1	8	1	32	96	-197,8180
NT-RUT-16-1-ot	{[(TiO2)2]16}1	16	1	32	96	-141,9680
NT-ANT-6-3-ot	{[(TiO2)4]6}3	6	3	72	216	-326,6390

NT-RUT-12-3-ot	{[(TiO2)2]12}3	12	3	72	216	-323,1400

The results in Table 4.8 for the different models above show that anatase nanotubes are more stable than rutile structures in terms of energy variation.

However, the difference between the energy variation values tended to decrease as the number of atoms in the structures increased. With values close to ΔE (eV).

5.1.4. Comparison of anatase nanotube models with the same number of atoms

Structures with the same number of atoms can be identified by the products of the n and m values. Example: The structures, NT-ANT-4-1- ot (n x m >> 4 x 1 = 4) and NT-ANT-2-2-ot (n x m >> 2 x 2 = 4), have 48 atoms each (See Table 4.1). Table 4.9 shows the anatase nanotube models used for the comparison:

Table 4.9 - Table of anatase nanotubes with the same number of atoms

ANATASE							
	Structure name	**Molecular Formula**	**(TiO2)4 (YZ)**	**(TiO2)4 (X)**	**(TiO2)4 Total**	**(Tiθ2) Total**	**Number of atoms**
1	NT-ANT-6-1-ot	{[(TiO2)4]6}1	6	1	6	24	72
1	NT-ANT-3-2-ot	{[(TiO2)4]3}2	3	2	6	24	72
1	NT-ANT-2-3-ot	{[(TiO2)4]2}3	2	3	6	24	72
2	NT-ANT-12-1-ot	{[(TiO2)4]12}1	12	1	12	48	144
2	NT-ANT-6-2-ot	{[(TiO2)4]6}2	6	2	12	48	144
2	NT-ANT-4-3-ot	{[(TiO2)4]4}3	4	3	12	48	144
3	NT-ANT-9-2-ot	{[(TiO2)4]9}2	9	2	18	72	216
3	NT-ANT-6-3-ot	{[(TiO2)4]6}3	6	3	18	72	216
4	NT-ANT-20-1-ot	{[(TiO2)4]20}1	20	1	20	80	240
4	NT-ANT-10-2-ot	{[(TiO2)4]10}2	10	2	20	80	240
5	NT-ANT-12-2-ot	{[(TiO2)4]12}2	12	2	24	96	288
5	NT-ANT-8-3-ot	{[(TiO2)4]8}3	8	3	24	96	288

Table 4.10 shows a comparison of the energy variation for structures with the same number of atoms.

Table 4.10 - Table of anatase nanotubes with the same number of atoms and their respective energy variations (eV)

	Structure name	Molecular Formula	Number of atoms	AE (eV)
1	NT-ANT-6-1-ot	{[(TiO2)4]6}1	72	-70,6697
1	NT-ANT-3-2-ot	{[(TiO2)4]3}2	72	-109,9690
1	NT-ANT-2-3-ot	{[(TiO2)4]2}3	72	-122,1100
2	NT-ANT-12-1-ot	{[(TiO2)4]12}1	144	-141,9680
2	NT-ANT-6-2-ot	{[(TiO2)4]6}2	144	-209,5750
2	NT-ANT-4-3-ot	{[(TiO2)4]4}3	144	-323,1400
3	NT-ANT-9-2-ot	{[(TiO2)4]9}2	216	-333,1950
3	NT-ANT-6-3-ot	{[(TiO2)4]6}3	216	-323,1400
4	NT-ANT-20-1-ot	{[(TiO2)4]20}1	240	-236,9220
4	NT-ANT-10-2-ot	{[(TiO2)4]10}2	240	-367,0590
5	NT-ANT-12-2-ot	{[(TiO2)4]12}2	288	-441,3190
5	NT-ANT-8-3-ot	{[(TiO2)4]8}3	288	-425,4480

The results in Table 4.10 show that, for the first comparisons (1) and (2), among the nanotubes with the fewest atoms, the longest structures, in this case the 3-level nanotubes (NT- ANT-2-3-ot and NT-ANT-4-3-ot), had the lowest energy variation values. The same occurs in comparison (4), where the 2-level nanotube with a smaller radius shows a smaller energy variation than a larger radius 1-level nanotube. Comparisons (3) and (5) showed very close values, in which the differences between the diameters did not lead to such a significant difference between the models. These results show that for the model proposed for anatase nanotubes, the length of the nanotubes (m levels) is a more influential factor than the diameter of the nanotube (n levels), between the models with the same number of atoms.

5.1.5. Calculation of winding energy

According to the methodology proposed in the study, the planes are first formed and then rolled up to form the nanotubes. Equation 4.3 shows the formation involved:

$$(mxn)[(TiO_2)_Y] \rightarrow \{[(TiO_2)_Y]_n\}_m \rightarrow \{[(TiO_2)_Y]_n\}_m \quad (4.3)$$

UNIDADE BÁSICA → PLANO → NANOTUBO

The values obtained for the energy variation between some isomers (planes and their respective nanotubes) can be seen in Table 4.11:

Table 4.11 - Table of isomers (nanotubes and planes) with their respective energy variations (eV) (PM7)

Nanotube	**AE (eV) Nanotube**	**Plan**	**AE (eV) Plan**	**AE (eV) Nanotube - AE (eV) Flat**
NT-ANT-2-1-ot	-26,5585	PL-ANT-2-1-ot1	-17,4605	-9,0980
NT-ANT-3-1-ot	-34,4746	PL-ANT-3-1-ot1	-29,2930	-5,1816
NT-ANT-11-1-ot	-130,1330	PL-ANT-11-1-ot1	-123,1000	-7,0330
NT-ANT-12-1-ot	-141,9680	PL-ANT-12-1-ot1	-134,9580	-7,0100
NT-ANT-19-1-ot	-225,0600	PL-ANT-19-1-ot1	-217,2710	-7,7890
NT-ANT-20-1-ot	-236,9220	PL-ANT-20-1-ot1	-228,9580	-7,9640
NT-ANT-2-2-ot	-73,9538	PL-ANT-2-2-ot1	-53,9381	-20,0157
NT-ANT-3-2-ot	-109,9690	PL-ANT-3-2-ot1	-88,8425	-21,1265
NT-ANT-10-2-ot	-367,0590	PL-ANT-10-2-ot1	-337,4570	-29,6020
NT-ANT-11-2-ot	-404,5160	PL-ANT-11-2-ot1	-374,6570	-29,8590
NT-ANT-12-2-ot	-441,3190	PL-ANT-12-2-ot1	-416,2120	-25,1070
NT-ANT-13-2-ot	-478,0890	PL-ANT-13-2-ot1	-458,5070	-19,5820
NT-ANT-14-2-ot	-514,8180	PL-ANT-14-2-ot1	-492,8510	-21,9670
NT-ANT-2-3-ot	-122,1100	PL-ANT-2-3-ot1	-88,7549	-33,3551
NT-ANT-3-3-ot	-182,9360	PL-ANT-3-3-ot1	-139,6780	-43,2580
NT-ANT-13-3-ot	-697,8670	PL-ANT-13-3-ot1	-677,1090	-20,7580
NT-RUT-6-1-ot	-46,5396	PL-RUT-6-1-ot1	-43,7147	-2,8249
NT-RUT-7-1-ot	-62,3395	PL-RUT-7-1-ot1	-53,6527	-8,6868
NT-RUT-13-1-ot	-114,0900	PL-RUT-13-1-ot1	-110,9760	-3,1140
NT-RUT-14-1-ot	-112,1130	PL-RUT-14-1-ot1	-111,3640	-0,7490
NT-RUT-19-1-ot	-162,9650	PL-RUT-19-1-ot1	-158,2990	-4,6660
NT-RUT-6-2-ot	-103,5880	PL-RUT-6-2-ot1	-99,3112	-4,2768
NT-RUT-9-2-ot	-163,4390	PL-RUT-9-2-ot1	-117,5490	-45,8900
NT-RUT-11-2-ot	-203,0530	PL-RUT-11-2-ot1	-197,0070	13,9540

NT-RUT-6-3-ot	-165,1120	PL-RUT-6-3-ot1	-161,7570	-3,3550

The results of the differences between the energy variations of nanotubes and pianos show that nanotubes tend to be more stable than pianos.

Figure 4.19 shows a graph of the average energy differences between the nanotube levels.

Figure 4.19 - Graph (Averages of ΔE (eV) differences X m levels)

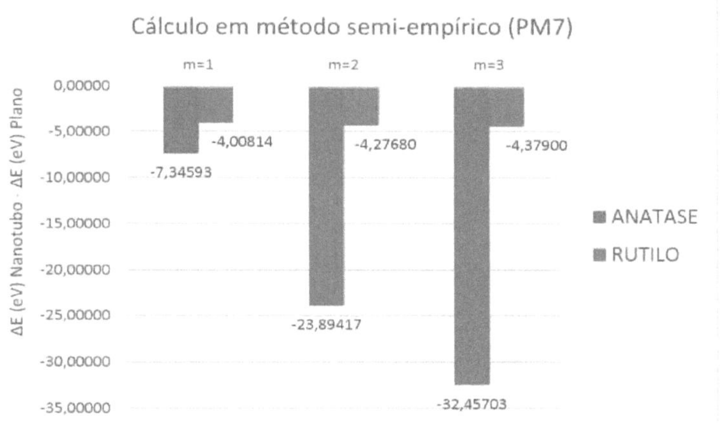

Relating these differences to the levels, it can be seen that for anatase nanotubes, the higher levels have a tendency to form nanotubes more easily. This behavior is no longer observed for rutile nanotubes. Although the 2^0 level of rutile nanotubes does not show consistent data, the U and 30 levels show similar results.

A possible indication of this difference between anatase and rutile is the basic unit of formation of the planes and nanotubes. This shows that anatase nanotubes tend to be more stable than rutile nanotubes.

5.1.6. Comparison of AE energy variation (eV) between semi-empirical (PM7) and ab initio methods.

Table 4.12 below compares the ΔE (eV) values of some of the anatase nanotube models analyzed.

Table 4.12 - Comparison between semi-empirical and *ab-initio* calculations

ΔE (eV)					
Structure	DFT/]	B3LYP	HF		PM7
	6-31G	6-311G	6-31G	6-311G	
NT-ANT-2-1-ot	-34,8674	-34,6987	-35,0023	-35,1008	-26,5585
NT-ANT-3-2-ot	-125,6752	-125,7489	-123,9780	-124,0187	-109,9690
NT-ANT-6-3-ot	-372,6210	-372,9514	-370,9832	-371,1296	-323,1400

Comparing the results between the same structure, the results between the 6-31G and 6-311G bases showed that there was no considerable difference, but between the *ab-initio* methods and the PM7 semi-empirical method, there was already a considerable difference in the results. This shows that there is a considerable influence of the other electrons in the system compared to the electrons in the valence layer.

5.2. HOMO-LUMO AND DENSITY OF STATES RESULTS FOR ANATASE AND RUTILE NANOTUBES

The HOMO-LUMO values determine the classification of materials as conductors, semiconductors or insulators, and the difference between the HOMO-LUMO values is called the *gap*. When the value is greater than 4.0 eV the material is classified as an insulator, between 1.0 eV and 4.0 eV it is classified as a semi-conductor, while below 1.0 eV it is classified as a conductor (FAHLMAN, 2011).

The values obtained for rutile and anatase nanotubes using the semi-empirical method (PM7) are shown in Table 6.12:

Table 4.13 - Table of HOMO, LUMO and gap values (eV) of anatase and rutile nanotubes obtained using the semi-empirical method (PM7)

Nanotube	HOMO (eV)	LUMO (eV)	*gap* (eV)
ANATASE NANOTUBES - 3 LEVELS			
NT-ANT-2-3-ot	-10,12	-2,20	7,92
NT-ANT-3-3-ot	-9,93	-2,35	7,59
NT-ANT-4-3-ot	-9,27	-3,15	6,12
NT-ANT-5-3-ot	-9,56	-2,50	7,07
NT-ANT-6-3-ot	-9,88	-2,88	7,00
NT-ANT-7-3-ot	-9,86	-2,59	7,27

NT-ANT-8-3-ot	-9,87	-2,60	7,27
NT-ANT-9-3-ot	-9,88	-2,61	7,27
NT-ANT-10-3-ot	-9,89	-2,62	7,27
NT-ANT-11-3-ot	-9,78	-2,90	6,88
NT-ANT-12-3-ot	-9,90	-2,63	7,27
NT-ANT-13-3-ot	-9,64	-2,76	6,87
NT-ANT-17-3-ot	-9,80	-2,94	6,85
NT-ANT-19-3-ot	-9,72	-2,87	6,85
NT-ANT-20-3-ot	-9,72	-2,87	6,85
ANATASE NANOTUBES - 2 LEVELS			
NT-ANT-2-2-ot	-9,56	-2,14	7,42
NT-ANT-3-2-ot	-10,05	-2,36	7,69
NT-ANT-4-2-ot	-9,90	-2,21	7,69
NT-ANT-5-2-ot	-10,11	-2,85	7,26
NT-ANT-6-2-ot	-10,15	-2,83	7,32
NT-ANT-7-2-ot	-10,14	-2,83	7,31
NT-ANT-8-2-ot	-10,13	-2,83	7,30
NT-ANT-9-2-ot	-9,69	-3,18	6,51
NT-ANT-10-2-ot	-9,74	-2,99	6,75
NT-ANT-11-2-ot	-9,38	-2,57	6,81
NT-ANT-12-2-ot	-9,42	-2,61	6,81
NT-ANT-13-2-ot	-9,45	-2,65	6,81
NT-ANT-14-2-ot	-9,48	-2,73	6,81
ANATASE NANOTUBES - 1 LEVEL			
NT-ANT-2-1-ot	-10,86	-2,40	8,46
NT-ANT-3-1-ot	-10,21	-3,00	7,22
NT-ANT-4-1-ot	-10,25	-2,98	7,26
NT-ANT-5-1-ot	-10,25	-2,98	7,27
NT-ANT-6-1-ot	-10,25	-2,98	7,27
NT-ANT-7-1-ot	-10,24	-2,97	7,27
NT-ANT-8-1-ot	-10,23	-2,96	7,27
NT-ANT-9-1-ot	-10,23	-2,97	7,26

NT-ANT-11-1-ot	-10,21	-2,92	7,29
NT-ANT-12-1-ot	-10,21	-2,85	7,36
NT-ANT-13-1-ot	-10,21	-2,85	7,36
NT-ANT-14-1-ot	-10,21	-2,85	7,37
NT-ANT-15-1-ot	-10,21	-2,84	7,37
NT-ANT-16-1-ot	-10,21	-2,84	7,37
NT-ANT-17-1-ot	-10,21	-2,84	7,37
RUTILO NANOTUBES - 3 LEVELS			
NT-RUT-6-3-ot	-6,55	-3,11	3,45
NT-RUT-8-3-ot	-6,62	-2,91	3,71
NT-RUT-9-3-ot	-6,37	-3,32	3,05
NT-RUT-10-3-ot	-6,31	-2,84	3,47
NT-RUT-11-3-ot	-6,52	-2,99	3,53
NT-RUT-12-3-ot	-6,61	-2,85	3,77
NT-RUT-13-3-ot	-6,12	-3,12	3,00
RUTILE NANOTUBES - 2 LEVELS			
NT-RUT-6-2-ot	-6,45	-2,82	3,63
NT-RUT-9-2-ot	-6,45	-2,72	3,73
NT-RUT-10-2-ot	-6,45	-2,87	3,58
NT-RUT-11-2-ot	-6,36	-3,33	3,02
NT-RUT-12-2-ot	-6,30	-3,25	3,04
NT-RUT-13-2-ot	-6,74	-2,76	3,99
NT-RUT-16-2-ot	-6,42	-3,05	3,37
NT-RUT-17-2-ot	-6,70	-3,12	3,58
NT-RUT-20-2-ot	-6,80	-3,34	3,46
RUTILE NANOTUBES - 1 LEVEL			
NT-RUT-6-1-ot	-6,28	-2,14	4,14
NT-RUT-7-1-ot	-6,48	-2,11	4,37
NT-RUT-8-1-ot	-6,72	-2,35	4,37
NT-RUT-9-1-ot	-6,20	-2,70	3,51
NT-RUT-10-1-ot	-6,39	-2,49	3,90
NT-RUT-11-1-ot	-6,05	-2,91	3,15

NT-RUT-12-1-ot	-6,40	-2,90	3,50
NT-RUT-13-1-ot	-6,31	-2,61	3,71
NT-RUT-14-1-ot	-6,41	-3,41	3,00
NT-RUT-15-1-ot	-6,58	-3,27	3,31
NT-RUT-16-1-ot	-6,19	-3,04	3,15
NT-RUT-17-1-ot	-6,59	-2,67	3,92
NT-RUT-18-1-ot	-6,53	-3,79	2,74
NT-RUT-19-1-ot	-6,24	-3,79	2,45
NT-RUT-20-1-ot	-6,13	-3,51	2,62

The graphs in Figure 4.20 and 4.21 below show the behavior of the *gap* values found in anatase and rutile nanotubes with their respective levels:

Figura 4.20 - Graphs of the *gap* values (ev) of the 3-level anatase nanotubes, obtained using the semi-empirical method (PM7)

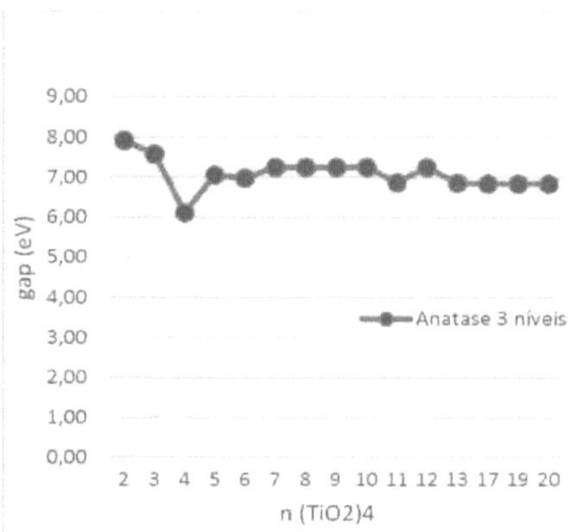

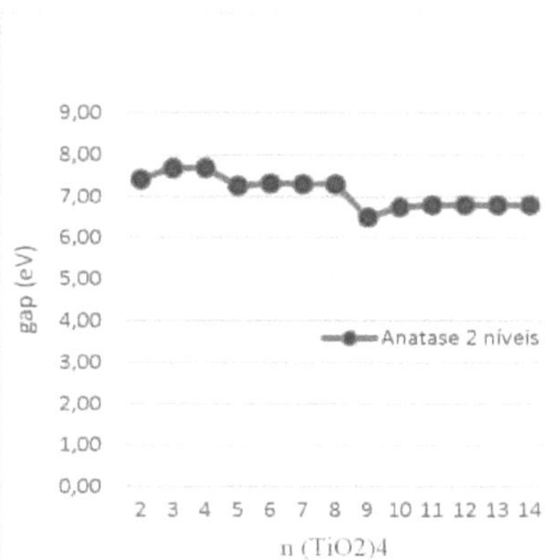

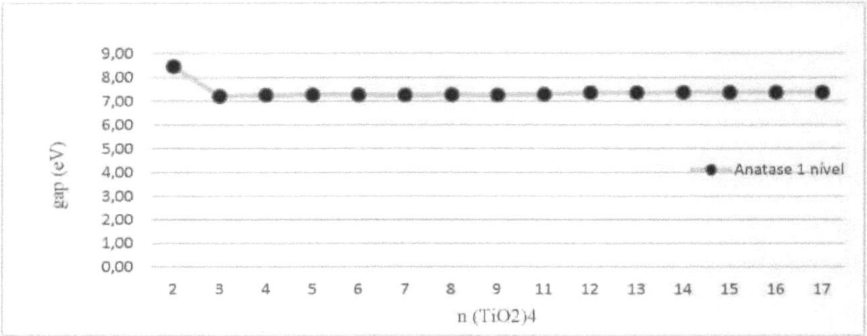

The results for the anatase nanotubes show that increasing the length and diameter do not contribute significantly to changing the *gap* values, without altering their conductivity properties. The anatase nanotubes maintained gap values in the range between 6.00 eV and 8.00 eV, behaving like an insulating material.

Figura 4.21 - Graphs of the gap values (eV) of the 3-level rutile nanotubes, obtained using the semi-empirical method (PM7)

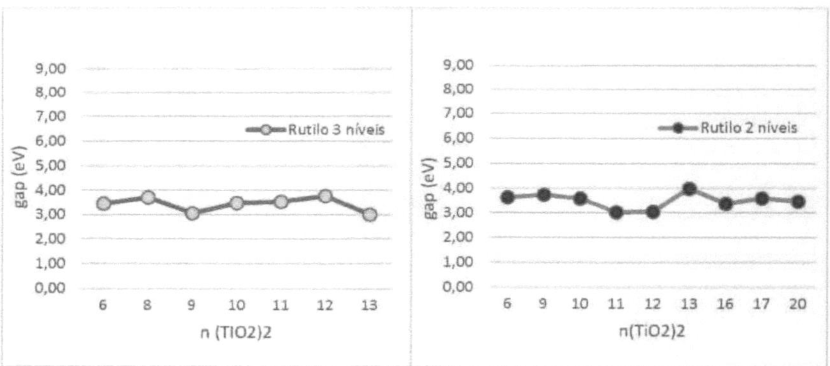

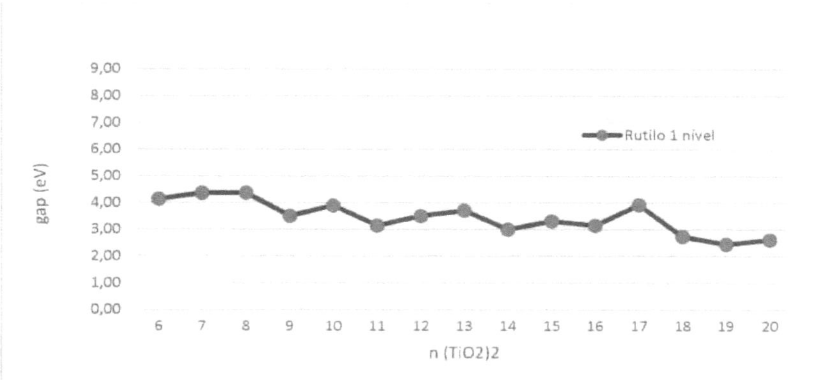

The results of the rutile nanotubes show that increasing the length and diameter do not contribute significantly to changing the gap values, as do the anatase nanotubes, without altering the conductivity properties. The rutile nanotubes maintained gap values in the range between 2.00 eV and 4.00 eV, classifying them as a semiconductor material.

DOS graphs allow visualization of the valence and conduction bands. The higher the value of the density of states at a given energy level, indicates the availability of states to be occupied at that level, while the cancellation of its value indicates the absence of states available to be occupied at the given energy level (PETROFF, *et al.* 1971).

The DOS of the smallest and largest diameter and length nanotubes of both anatase and rutile was analyzed. The DOS graphs are shown in Figures 4.22 to 4.29.

Figure 4.22 - DOS - NT-ANT-20-3-ot (PM7)

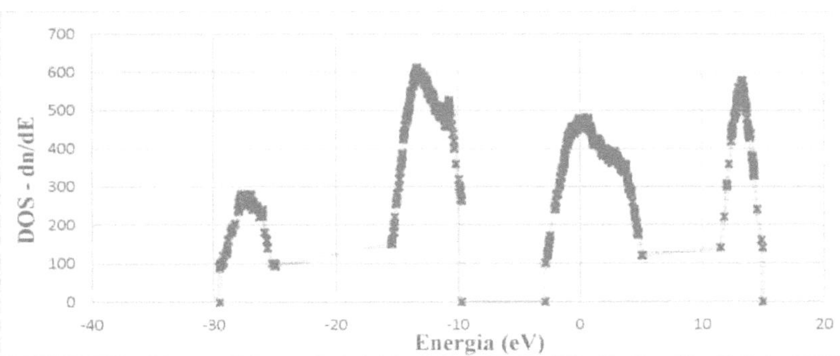

Figure 4.23 - DOS - NT-ANT-2-3-ot (PM7)

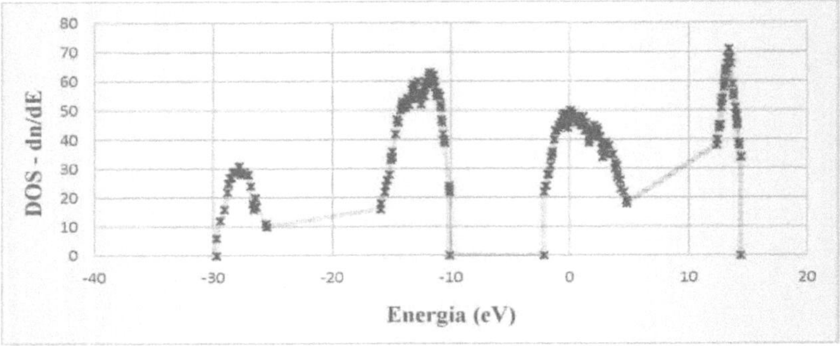

Figure 4.24 - DOS - NT-ANT-14-2-ot (PM7)

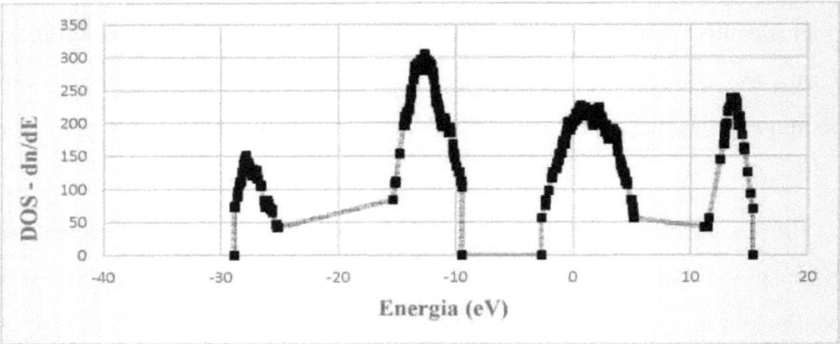

Figure 4.25 - DOS - NT-ANT-2-2-ot (PM7)

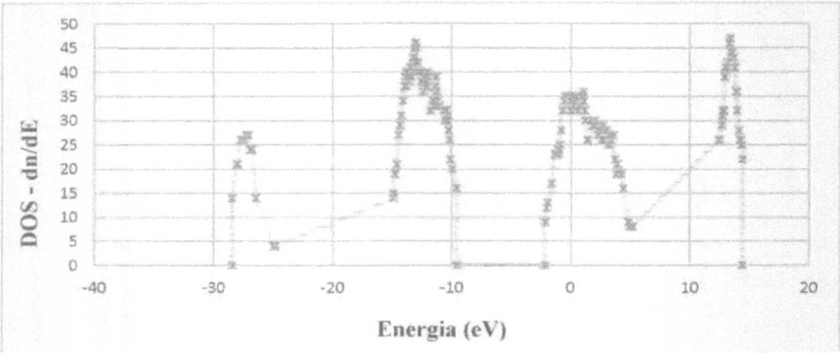

Figure 4.26 - DOS - NT-RUT-13-3-ot (PM7)

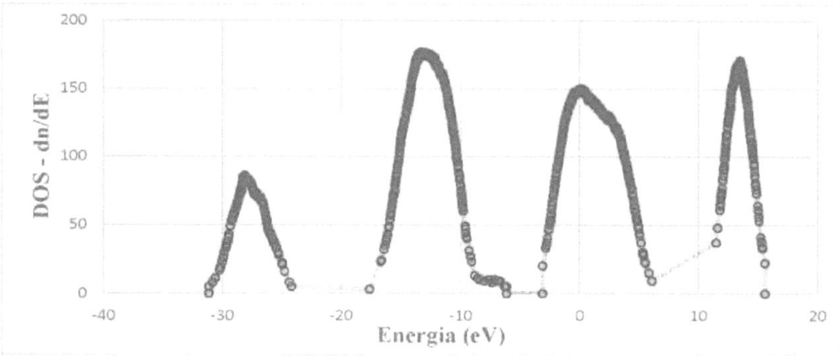

Figure 4.27 - DOS - NT-ANT-6-3-ot (PM7)

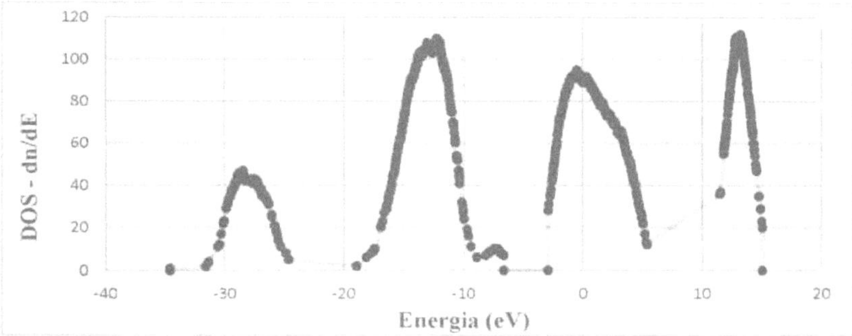

Figure 4.28 - DOS - NT-ANT-20-2-ot (PM7)

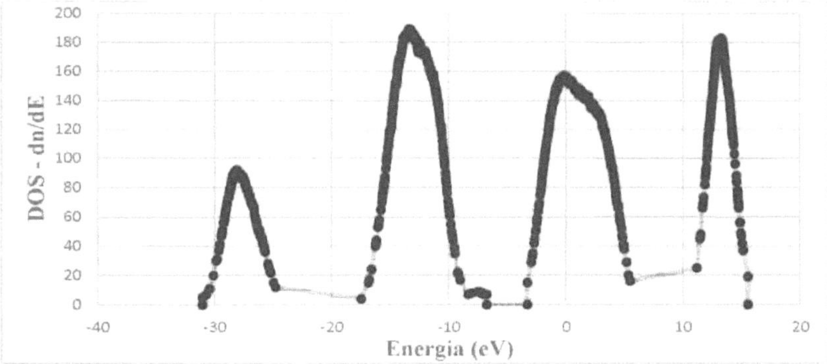

Figure 4.29 - DOS - NT-ANT-6-2-ot (PM7)

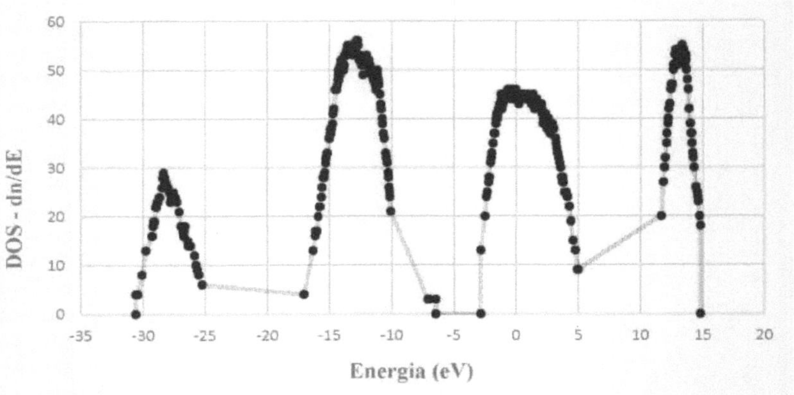

The DOS results show that the valence band of the anatase nanotubes is concentrated in the -15.00 eV to -10.00 eV range, *while* the conduction band has values in the -2.50 eV to 5.00 eV range. In the case of rutile nanotubes, the DOS results show that the valence band was concentrated in the range between -16.00 eV to -6.00 eV, while the conduction band showed values in the range between -3.00 eV to 5.00 eV.

Gap results calculated by ab-initio methods were also collected from anatase structures. These results are described in Table 4.14.

Table 4.14 - Table of *gap* results (eV) in ab-initio methods

gap **(eV)**					
Structure	**DFT/B3LYP**		**HF**		**Semi-Empirical**
	6-31G	6-311G	6-31G	6-311G	PM7
NT-ANT-2-1-ot	2,47	2,46	9,43	9,70	8,46
NT-ANT-3-2-ot	2,03	2,05	8,95	-	7,69
NT-ANT-6-3-ot	2,29	2,29	9,28	9,48	7,00

The results showed a big difference between the results of the HF and DFT methods. The HF method overestimated the *gap* values compared to the semi-empirical method, while the DFT method values were lower compared to the semi-empirical and HF methods. When comparing the 6-31G and 6-311G bases, there was no considerable variation between their values. In comparison with experimental and theoretical results found in the literature, the values found by the DFT method, with values between 2.00

eV and 2.50 eV, are the closest (see table 2.2).

5.3. RESULTS OF THE DISTANCES AND DIAMETERS OF ANATASE AND RUTILE NANOTUBES

The distances between the atoms of the optimized rutile and anatase nanotubes were calculated using the arithmetic mean of the bond distances of the **Ti-Ti**, **Ti-O** and **O-O** ligands, as well as the diameter values in the middle and at the ends of the optimized nanotubes.

In order to compare the distances of the **Ti-Ti**, **Ti-O** and **O-O** ligands and the diameters found between the models obtained, we selected models from the 2^0 and 3^0 levels, collecting 03 nanotubes from each level of both types (anatase and rutile), collecting nanotubes with smaller diameters, intermediate diameters and larger diameters. In this case, the structures were selected:

- Anatase: NT-ANT-2-2-ot, NT-ANT-9-2-ot, NT-ANT-14-2-ot, NT- ANT-2-3-ot, NT-ANT-10-3-ot, NT-ANT-20-3-ot.
- Rutile: NT-RUT-6-2-ot, NT-RUT-10-2-ot, NT-RUT-20,2-ot, NT- RUT-6-3-ot, NT-RUT-8-3-ot, NT-RUT-13-3-ot.

Table 4.15 shows the results of the distances between the ligands that make up the nanotubes.

Table 4.15 - Average distances of anatase and rutile nanotubes

ANATASE					
NT-ANT-2-2-ot		**NT-ANT-9-2-ot**		**NT-ANT-14-2-ot**	
Average connection distances (Â)					
Ti-Ti	3,33	Ti-Ti	3,32	Ti-Ti	3,31
Ti-O	2,17	Ti-O	1,93	Ti-O	1,89
O-O	2,96	O-O	3,00	O-O	3,00
NT-ANT-2-3-ot		**NT-ANT-10-3-ot**		**NT-ANT-20-3-ot**	
Average connection distances (Â)					
Ti-Ti	3,37	Ti-Ti	3,23	Ti-Ti	3,29

Ti-O	1,92	Ti-O	2,03	Ti-O	2,17
O-O	2,99	O-O	2,99	O-O	3,00
RUTILO					
NT-RUT-6-2-ot		**NT-RUT-10-2-ot**		**NT-RUT-20-2-ot**	
Average connection distances (Â)					
Ti-Ti	3,36	Ti-Ti	3,35	Ti-Ti	3,37
Ti-O	2,20	Ti-O	2,15	Ti-O	2,14
O-O	2,95	O-O	2,97	O-O	2,99
NT-RUT-6-3-ot		**NT-RUT-8-3-ot**		**NT-RUT-13-3-ot**	
Average connection distances (Â)					
Ti-Ti	3,38	Ti-Ti	3,35	Ti-Ti	3,36
Ti-O	2,20	Ti-O	2,28	Ti-O	2,21
O-O	2,93	O-O	2,94	O-O	2,97

Table 4.15 shows the average distances of the optimized models obtained:

- The average distances of the **Ti-Ti** and **Ti-O** ligands of anatase nanotubes are smaller compared to rutile nanotubes;
- The average **O-O** bond distances of anatase nanotubes are greater than those of rutile nanotubes;

Table 4.16 describes the selected model diameters of anatase and rutile nanotubes.

Table 4.16 - Table of anatase and rutile nanotube diameters

		Nanotube diameter (nm)		
Structure	**Molecular Formula**	**End Initial**	**End Final**	**Medium**
NT-ANT-2-2-ot	[[(TiO2)4]2]2	0,82	0,91	-
NT-ANT-9-2-ot	[[(TiO2)4]9]2	2,93	3,15	-
NT-ANT-14-2-ot	[[(TiO2)4]14]2	4,37	4,97	-
NT-ANT-2-3-ot	[[(TiO2)4]2]3	0,97	0,68	0,81
NT-ANT-10-3-ot	[[(TiO2)4]10]3	3,86	3,77	3,80
NT-ANT-20-3-ot	[[(TiO2)4]20]3	7,44	7,18	7,34
NT-RUT-6-2-ot	[[(TiO2)2]6]2	0,86	0,57	-
NT-RUT-10-2-ot	[[(TiO2)2]10]2	0,91	1,20	-

NT-RUT-20-2-ot	[[(TiO2)2]20]2	1,75	1,54	-
NT-RUT-6-3-ot	[[(TiO2)2]6]3	0,81	0,87	0,75
NT-RUT-8-3-ot	[[(TiO2)2]8]3	0,92	0,72	0,50
NT-RUT-13-3-ot	[[(TiO2)2]13]3	1,08	1,00	0,89

5.4. RESULTS OF THE INTERACTION OF THE ANATASE NANOTUBE (NT-ANT-10-03-ot) WITH GASES

In the study of the interaction of gases with titanium dioxide nanotubes, we simulated the interaction of an anatase nanotube, in this case the NT- ANT-10-03-ot nanotube, with the diatomic molecules CO and H_2. In order to evaluate these interactions, an analysis was carried out using the results of the adsorption distances between the nanotube and the carbon monoxide and hydrogen gas molecules. The variations in adsorption energy and the gap of the structures after interaction were also analyzed.

Equations 4.4 and 4.5 were used to evaluate the variation in adsorption energy between the nanotube and the molecules:

$$\Delta E_{ads\text{-}H_2} = E_{total\,(NT\text{---}nH_2)} - E_{total(NT)} - nE_{total(H_2)} \qquad (4.4)$$

$$\Delta E_{ads\text{-}CO} = E_{total\,(NT\text{---}nCO)} - E_{total(NT)} - nE_{total(CO)} \qquad (4.5)$$

In Equations 4.4 and 4.5, the energy variation is calculated as the difference between the total energy of the system (nanotube interacting with the molecule), the total energy of the isolated nanotube and the energy of the isolated molecule.

5.4.1. Anatase nanotube interaction with H_2

The theoretical simulations of the interaction between the NT-ANT-10-3-ot nanotube and the H_2 molecule took place with the atoms of the TiO2 units located at the ends and in the middle of the nanotubes, varying the number of molecules interacting with the nanotube. In the simulation of the nanotube in the presence of H_2, there was no interaction, *since* the distances between the nanotube and the H2 molecules were between 3.90 and 4.50. Figure 4.30 illustrates this situation.

Figure 4.30 - TiO nanotube$_2$ with H_2

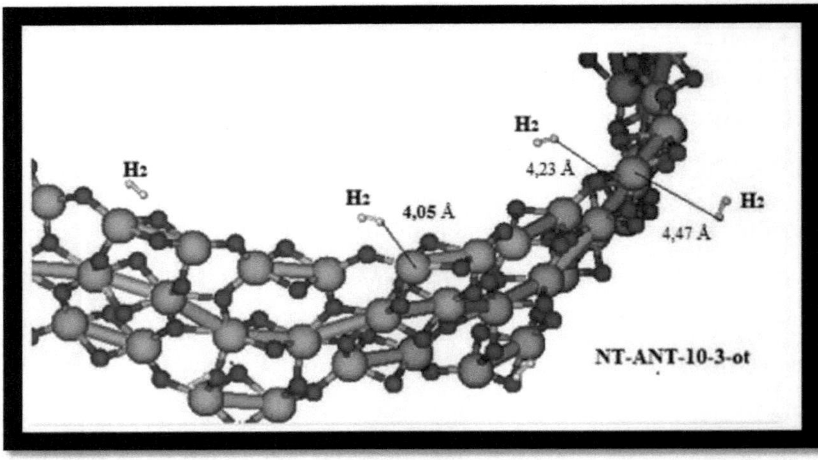

Source: Author

These simulations gave the following results, which are described in Table 4.17.

Table 4.17 - Table of ΔEads and gap results presented in the simulation of NT-ANT-10-3 in the presence of hydrogen (PM7)

Interaçâo [NT-ANT-10-3-ot----⅛]			
Number of H2 molecules	1	5	20
ΔEads (eV)	-0,00529	-0,04181	-0,25601
gap (eV)	7,27	7,27	7,28

The results shown in Table 4.17 show that there were no significant changes in the energy and *gap* values of the structure, which remained the same as the *gap* value of the isolated structure (7.27 eV).

The fact that there is no interaction between "pure" titanium dioxide nanotubes and the gas H_2 is reported by Galstyan in his experimental apparatus, where according to the author, doping with Nb increases the performance of the TiO nanotube sensor$_2$ (GALSTYAN, *et al.* 2013). Li and his colleagues found that anatase nanotubes doped with Ni increased the sensitivity of the sensor in an atmosphere of H_2 , which was verified experimentally and also by means of DFT calculations on a PBE (*Perdew-Burke-Ernzerhof*) basis. Li also found a decrease in the structure's gap from 2.40 eV to 0.26 eV (LI, *et al.* 2014).

Simulations of the NT-ANT-10-3-ot nanotube doped with Pt and Au were carried out and showed that the H2 molecule interacted with the Au atom, changing the nanotube *gap* from 7.27 eV to 5.56 eV. The distances found between H2 and the nanotube, which were around 4.00 Â, became closer with doping, forming interactions with distances between 2.03 Â ά 2.31 Â. Figure 4.31 illustrates the results obtained with the doping of Pt and Au in the NT-ANT-10-3-ot structure.

Figure 4.31 - Anatase nanotube doped with Pt and Au interacting with H2 gas.

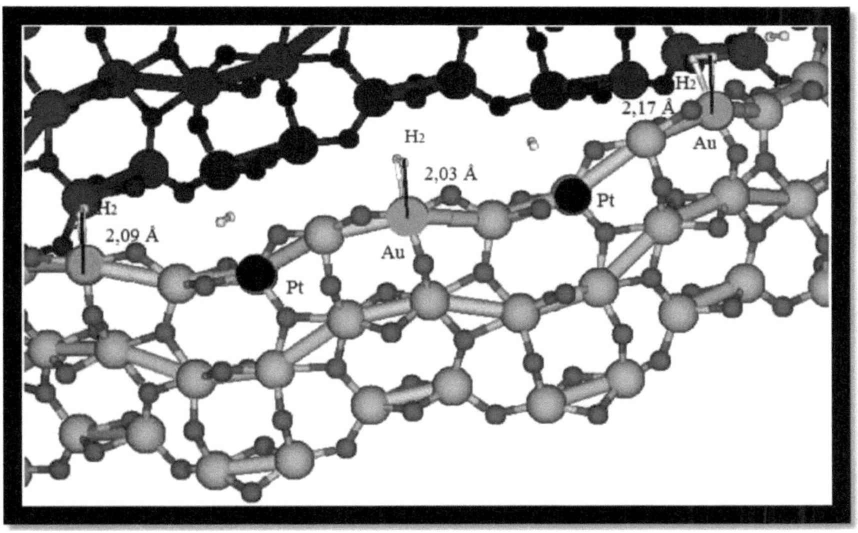

Source: Author

5.4.2. Interaction of the anatase nanotube with CO

The theoretical simulations of the interaction between the NT-ANT-10-3-ot nanotube and the CO molecule took place with the atoms of the TiO2 units located at the ends and in the middle of the nanotubes, varying the number of molecules interacting with the nanotube. The simulation of the nanotube in the presence of CO showed the formation of new bonds between the nanotube and the carbon monoxide molecules, a fact that can be proven by the distance between the oxygen atoms of the monoxide and the Ti atom, which showed values between 1,20 Â to 1.40 Â, while the distances between the Ti-O in the nanotubes range from 1.90 Â to 2.25 Â. This is not just an adsorption resulting from a physical interaction, but the formation of a chemical bond, involving the covering of orbitals. Figure 4.32 illustrates this situation.

Figure 4.32 - TiO2 nanotube with CO

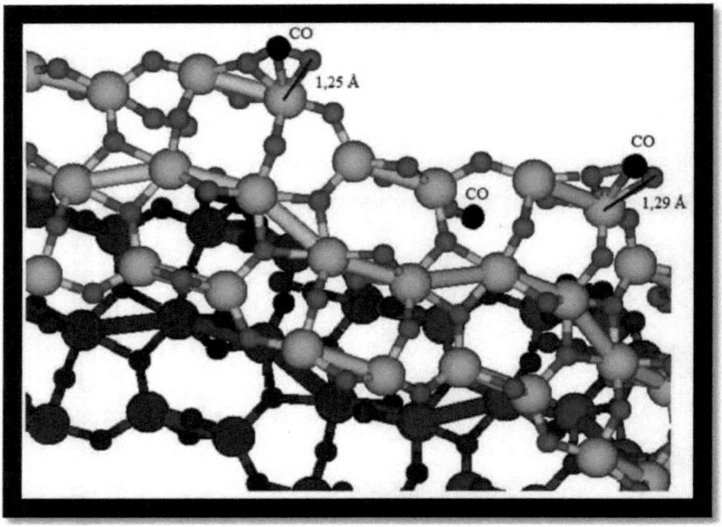

Source: Author

These simulations gave the following results for the gap and adsorption energy variation, described in Table 4.18.

Table 4.18 - Table of ΔEads and gap results presented in the NT-ANT-10-3 simulation in the presence of carbon monoxide (PM7)

NT-AN interaction	**ΓΓ-10-3-otCO**		
Number of CO	1	10	20
Energy variation (eV)	-0,381240	-8,22271	-16,51785
gap (eV)	7,20	7,00	7,29

The results shown in Table 4.18 show that there were no significant changes in the gap values of the structure, which remained in the same gap value ranges as the isolated structure. However, the values obtained for the energy variations show that the structure has acquired greater stability compared to the isolated structure.

6. CONCLUSIONS

- Analysis of the variation in energy with the 3 levels of anatase nanotubes showed that the trend towards stability is linked to the increase in diameter and consequently the decrease in curvature;
- The model with the lowest energy variation was NT-ANT-20-3-ot (ΔE = -1037.4700 eV);
- The same behavior was verified for rutile nanotubes, also demonstrating that the stability trend for their models is a function of diameter and, consequently, the decrease in curvature.
- The model with the lowest energy variation was NT-RUT-13-3-ot (ΔE = -349.85100 eV);
- By relating ΔE to the number of levels n, it can be seen that there is a linear behavior between the models obtained. However, it cannot be said that this behavior can be described for nanotubes with n levels above 20, as there is a limit to the diameter of the nanotubes;
- The comparison data between the rutile and anatase models showed that the anatase models were more stable than the rutile models;
- Comparing anatase models with the same number of atoms, but with different n and m levels, it was found that the m level, which is related to length, is a more influential factor than the n level (diameter), among the anatase nanotube models obtained;
- The winding energy calculations between the isomers (planes and nanotubes) showed that the nanotube conformation is more stable than the planes, for both rutile and anatase.
- The energy variation results of the *ab-initio* methods were not close to the results presented by the semi-empirical method;
- There were no discrepant results between the bases used in the study, 6-31G and

6-311G;

- The rutile nanotubes showed *gap* values of around 2.00 to 4.00 eV, which are characteristic of semiconductors, while the anatase nanotubes showed values in the range of 6.00 eV to 8.00 eV for the PM7 calculation;

- The results of the *gap* calculations in anatase nanotubes from the DFT method (2.00 eV to 2.50 eV) underestimated the values found in the semi-empirical method, while the results from the HF method overestimated these values. Theoretical and experimental results found in the literature show greater proximity to the results found in the DFT method;

- The average distances of the **Ti-Ti** and **Ti-O** ligands in anatase nanotubes are smaller compared to rutile nanotubes;

- The average **O-O** bond distances of anatase nanotubes are greater than those of rutile nanotubes;

- Adsorption between an anatase nanotube and the H2 molecule can be simulated by doping the structure with Au.

- The results of the interaction with the CO molecule showed the formation of a chemical bond between the CO molecule and the NT- ANT-10-3-ot structure;

- The methodology proposed, as well as the models created, have the potential to be applied to the simulation of titanium dioxide planes and nanotubes.

7. BIBLIOGRAPHICAL REFERENCES

BAVYKIN, D. V.; PARMON, V. N.; LAPKIN, A. A.; & WALSH, F. C. The effect of hydrothermal conditions on the mesoporous structure of TiO_2 nanotubes. Journal of Materials Chemistry, v. 14, n. 22, p. 3370-3377, 2004.

BELTRAN, A.; ANDRES, J.; SAMBRANO, J. R.; & LONGO, E. Density Functional Theory Study on the Structural and Electronic Properties of Low Index Rutile Surfaces for TiO /SnO /TiO_{222} and SnO /TiO /SnO_{222} Composite Systems. The Journal of Physical Chemistry A, v. 112, n. 38, p. 8943-8952, 2008.

BERNSTEIN, H. J. Home page for Rasmol or OpenRasmol. Available at: http://www.rasmol.org/. Accessed on July 12, 2014.

BETHUNE, D. S.; KIANG, C. H.; DEVRIES, M. S.; GORMAN, G.; SAVOY, R.; VAZQUEZ, J.; BEYERS, R. Cobalt-catalyzed growth of carbon nanotubes with single-atomic-layer walls. Nature, v.363, n.17, p.603-607, 1993.

BOLINK, H. J.; CORONADO, E.; OROZCO, J.; SESSOLO, M. Efficient polymer light-emitting diode using air-stable metal oxides as electrodes. *Adv. Mater. 21*, p. 79-82, 2009.

CARVALHO, A C. M. Electronic Structure of Carbon Nitride Nanotubes. 2004. 167 f. Thesis (Doctorate in Science) - "Gleb Wataghin" Institute of Physics, State University of Campinas, Sao Paulo, 2004.

COSTA, L. Application of titania nanotubes in the photodegradation of dyes. PhD in Chemistry - Institute of Chemistry, University of Brasilia, Brasilia, 2009.

COLLINS, P. G; AUVORIS, P. Nanotubes for electronics, Scientific American Magazine; p. 62 - 69, 2000.

DEWAR, M.J.S.; ZOEBISCH, E.F.; HEALY, E.F.; STEWART, J.J.P. J. Am. Chem. Soc. 107, 3902, 1995.

DIEBOLD, U.; RUZYCKI, N.; HERMAN, G. S.; & SELLONI, A. One step towards bridging the materials gap: surface studies of TiO2 anatase. Catalysis Today, v. 85, n.

2, p. 93-100, 2003.

EGGINS, Brian R. Chemical sensors and biosensors. University of Ulster, Northern Ireland. John Wiley & Sons, p. 1-9, 2008.

FERNANDES, Jesum Alves. Sensitization of TiO2 nanotubes with CdSe by RF magnetron sputtering for application in photoelectrochemical cells. PhD thesis, UFRGS-RS, Rio Grande do Sul 2014.

FAHLMAN, B. D.; Materials Chemistry, 2 ed. Springer Science & Business Media, Netherlands, p. 239, 2011.

FELDMAN, Y.; WAASSERMAN, E.; SROLOVITZ, D. J.; & TENNE, R. High-rate, gas-phase growth of MoS2 nested inorganic fullerenes and nanotubes. Science, v. 267, n. 5195, p. 222-225, 1995.

GALSTYAN, V.; COMINI, E.; FAGLIA, G.; & SBERVEGLIERI, G. TiO2 Nanotubes: Recent Advances in Synthesis and Gas Sensing Properties. Sensors, v. 13, n. 11, p. 14813-14838, 2013.

GALSTYAN, V.; COMIMI, E.; FAGLIA, G.; VOMIERO, A.; BORGESE, L.; BONTEMPI, E.; SBERVEGLIERI, G. Fabrication and investigation of gas sensing properties of Nb-doped TiO2 nanotubular arrays. *Nanotechnology,* 2012.

GANGULY, D.; ZHAO, Y. Fabrication of TiO2 Nanotube Arrays Using 1, 2-Propanediol Electrolyte for Application in Biomedical Implants. In: TMS 2014 143rd Annual Meeting & Exhibition, Annual Meeting Supplemental Proceedings. John Wiley & Sons, p.191, 2014.

GAUSSIAN, version 03. Available at: http://www.gaussian.com/. Accessed on: July 26, 2014.

GNUPLOT version 4.6.5 releesed. Available at: http://www.gnuplot.info/. Accessed on: July 12, 2014.

GONÇALVES, E. M. S. Intelligent and reinforced scaffolds for bone regenerative medicine. Master's dissertation - University of Aveiro, 2012.

GONG, X. Q.; SELLONI, A.; BATZIL, M.; & DIEBOLD, U. Steps on anatase TiO_2 (101). Nature materials, v. 5, n. 8, p. 665-670, 2006.

GUO, Z.; PREZHDO, O. V.; HOU, T.; CHEN, X.; LEE, S. T.; & LI, Y Fast Energy Relaxation by Trap States Decreases Electron Mobility in TiO2 Nanotubes: Time-Domain Ab Initio Analysis. The Journal of Physical Chemistry Letters, 2014.

HACOHEN, Y. R.; GRUBAUN E.; TENNE, R.; SLOAN, J.; HUTCHISON, J. L. Cage structures and nanotubes of NiCl2.Nature, v. 395, n. 6700, p. 336337, 1998.

HAMERS, B.; ST, PGA J.; VELD, M. A. J. The Wondrous World of Carbon Nanotubes. pp. 004, 2003.

HARRIS, P. J. F. Carbon nanotube science: synthesis, properties and applications. Cambridge University Press, pp. 003-006, 2009.

HOHENBERG, P.; KOHN, W. "Inhomogeneous Electron Gas". American Physical Society. v.136, pp. 864-871, 1964.

HOYER, Patrick. Formation of a titanium dioxide nanotube array. Langmuir, v. 12, n. 6, p. 1411-1413, 1996.

HOWARD, C. J.; SABINE, T. M.; DICKSON, FIONA. Structural and thermal parameters for rutile and anatase. Acta Crystallographica Section B: Structural Science, v. 47, n. 4, p. 462-468, 1991.

HUTLEEN, J. C.; JIRAGE, K. B.; MARTIN, C. R. Introducing chemical transport selectivity into gold nanotubule membranes. Journal American Chemical Society, v. 120, p. 6603-6604, 1998.

IIJIMA, S. Helical microtubules of graphitic carbon. Nature, v.354, n. 6348, p. 56-58, 1991.

KANG, T. S.; SMITH, A. P.; TAYLOR, B. E.; DURSTOCK, M. F. Fabrication of highly-ordered TiO2 nanotube arrays and their use in dye- sensitized solar cells. Nano Letters, v. 9(2), p. 601-606, 2009.

KASUGA, T.; HIRAMATSU, M.; HOSON, A.; SEKINO, T.; & NIIHARA, K.

Formation of titanium oxide nanotube. Langmuir, v. 14, n. 12, p. 31603163, 1998.

KIM, W. R.; PARK, H.; CHOI, W. Y. TiO2 micro-flowers composed of nanotubes and their application to dye-sensitized solar cells. **Nanoscale research letters**, v. 9, n. 1, p. 1-10, 2014.

KOHN, W.; SHAM, L.J. "Self-Consistent Equations Including Exchange and Correlation Effects". American Physical Society, v.140, n.04A, 1965.

KOWALSKI, P. M.; MEYER, B.; MARX, D. Composition, structure, and stability of the rutile TiO2 (110) surface: Oxygen depletion, hydroxylation, hydrogen migration, and water adsorption. Physical Review B, v. 79, n. 11, p. 115410, 2009.

LEITE, A. L. SORIANO, J. P.; SALES, G. A. A.; & Sales, W. F. Nanomaterials used as structural reinforcements in composite materials and their application in civil construction: an overview. Revista Interatividade, p. 39-44, 2014.

LEVINE, I.N. Quimica Cuàntica. 5ed. Brooklyn, New York, Prenticee Hall, p 275 - 676, 2001.

LI, J.; ZHOU, H.; QIAN, S.; LIU, Z.; FENG, J.; JIN, P.; & LIU, X. Plasmonic gold nanoparticles modified titania nanotubes for antibacterial application. Applied Physics Letters, v. 104, n. 26, p. 261110, 2014.

LI, Z.; DING, D.; LIU, Q.; NING, C.; & WANG, X. Ni-doped TiO2 nanotubes for wide-range hydrogen sensing. Nanoscale research letters, v. 9, n. 1, p. 118, 2014.

LIU, Z.; CAO, Z.; DENG, B.; WANG, Y.; SHAO, J.; KUMAR, P.; & CHENG, G. J. Ultrafast and scalable laser liquid synthesis of tin oxide nanotubes and its application in lithium ion batteries. Nanoscale, v. 6, n. 11, p. 5853-5858, 2014.

LIU, W.; SU, P.; WANG, N.; MA, Y.; LIU, Y.; WANG, J. & WEBSTER, T. J.; The synthesis of TiO_2 nanotubes with ZnO nanoparticles to possess antibacterial properties with no stem cell toxicity. Nanoscale, 2014.

LOSILLA, J. A.; RATANATAWANATE, C.; BALKUS J. R.; KENNETH J. Synthesis of TiO_2 nanotube films via pulsed laser deposition followed by a hydrothermal treatment. Journal of Experimental Nanoscience, v. 9, n. 2, p. 126-137, 2014.

MACAK, J. M.; TSUCHIYA, H.; GHICOV, A.; YASUDA, K.; HAHN, R.; BAUER, S.; & SCHMUKI, P. TiO2 nanotubes: Self-organized electrochemical formation, properties and applications. Current Opinion in Solid State & Materials Science, v. 11, p. 3-18, 2007.

MAIYALAGAN, T.; VISWANATHAN, B.; VARADARAJU, U. V. Fabrication and characterization of uniform TiO2 nanotube arrays by sol-gel template method. Bulletin of Materials Science, v. 29, n. 7, p. 705, 2006.

MOELLMANN, J.; EHRLICH, S.; TONNER, R.; & GRIMME, S. A DFT- D study of structural and energetic properties of TiO2 modifications. **Journal of Physics: Condensed Matter**, v. 24, n. 42, p. 424206, 2012.

MOLDEN, A pre-and post processing program of molecular and electronic structure. Available at: <http://www.cmbi.ru.nl/molden/. Accessed on July 12, 2014.

NAKAMURA, H.; MATSUI, Y.; Silica gel nanotubes obtained by the solgel method. Journal of the American Chemical Society, v. 117, n. 9, p. 26512652, 1995.

NUZZI, F.; ANGELIS, D. F.; DFT investigations of formic acid adsorption on single-wall TiO_2 nanotubes: effect of the surface curvature.The Journal of Physical Chemistry C, v. 115, n. 5, p. 2179-2186, 2010.

O'CONNELL, M. J. Carbon nanotubes: properties and applications. CRC press, 2012.

PARR, R G.; YANG, W. Density-Functional Theory of Atoms and Molecules. Oxford: Oxford University Press, 1989.

PARR, R.G.; CRAIG, D. P.; ROSS, I.G Molecular orbital calculations of the lower excited electronic levels of Benzene, configuration interaction included. Journal of Chemical Physics, n.12, v.18, pp.1561-1564, 1950.

MAJEWSKI, P. Aniodic TiO_2 Nanotubes Synthesis and Applications.Recent Patents on Materials Science, v. 7, n. 1, p. 1-7, 2014.

PETROFF, I.; VISWANATHAN, C. R. Calculation of Density of States in W, Ta, and Mo. Electronic density of states: Based on invited and contributed papers and discussion, v. 2, n. 323, p. 53, 1971.

RAMAMOORTHY, M.; VANDERBILT, D.; KING-SMITH, R. D. First- principles calculations of the energetics of stoichiometric TiO2 surfaces. Physical Review B, v. 49, n. 23, p. 16721, 1994.

ROCHA, G.B.; FREIRE, R.O.; SIMAS, A.; STEWART, J.J.P.J. Comp. Chem., 27, 1101, 2006.

SANTOS, J D. A Quàntica Quimica Aplicada ao Buckminsterfulereno C60. 2000. fls 24 -32 PhD in Science - Fisico-Quimica Departamento de Quimica, Universidade Federal de Sao Carlos, Sao Paulo-SP, 2000).

SANTOS, C.A, Feynman, o profeta da nanotecnologia, Revista CH Instituto Ciência Hoje, Universidade Federal do Rio Grande do Sul, 2007.

SCHWEICHER, J.; DESAI, T. A. Facile synthesis of robust free-standing TiO_2 nanotubular membranes for biofiltration applications. Journal of applied electrochemistry, v. 44, n. 3, p. 411-418, 2014.

SCHMUKI, P.; ROY, P.; KIM, D.; LEE, K.; SPIECKER, E. *et al.* TiO_2 nanotubes and their application in dye-sensitized solar cells. Nanoscale, v. 2, n. 1, p. 45-59, 2010.

SHKLOVER, V.; NAZEERUDDIN, M. K.; ZAKEERUDDIN, S. M.; BARBE, C.; KAY, A.; HAIBACH, T.; & GRATZEL, M. Structure of nanocrystalline TiO_2 powders and precursor to their highly efficient photosensitizer. Chemistry of materials, v. 9, n. 2, p. 430-439, 1997.

SILVA, J. R. G.; AZEVEDO, R. T.; SILVA, F. A. The importance of gas detection to prevent damage to safety, the environment and health: sources of interference in signals 4 to 20 Ma. Bolsista de Valor, v. 1, n. 1, p. 359-364, 2010.

SONG, T.; HAN, H.; CHOI, H.; LEE, J. W.; PARK, H., LEE, S., ... & PAIK, U. TiO2 nanotube branched tree on a carbon nanofiber nanostructure as an anode for high energy and power lithium ion batteries. Nano Research, v. 7, n. 4, p. 1-11, 2014.

SOUZA, F. M. Structural analysis and synthesis of titanium dioxide nanotubes. 2011. Dissertation (Master's Degree in Materials Science) - Postgraduate Program in Materials for Engineering, Universidade Federal de Itajubà, Itajubà, Minas Gerais, 2011.

STEWART, J. J. P. Stewart Computational Chemistry. MOPAC2012. Available at: http://openmopac.net, Accessed on: July 26, 2014.

STEWART, J. J. P. Mol. Model. 13, 1173, 2007.

SZABO, A. E. OSTLUND, N. S. Modern Quantum Chemistry. Mineola, New York: Dover Publications, INC., p. 40 - 47, 2006.

TENNE, R.; MARGULIS, L.; GENUT, M.; HODES, G. Polyhedral and cylindrical structures of tungsten disulphide. Nature, v. 360, n. 6403, p. 444446, 1992.

THOMAS, A. G.; FLAVELL, W. R.; MALLICK, A. K.; KUMARASINGHE, A. R.; TSOUTSOU, D.; KHAN, N.; WIAME, F. *et al.* Comparison of the electronic structure of anatase and rutile TiO_2 singlecrystal surfaces using resonant photoemission and X-ray absorption spectroscopy. Physical Review B, v. 75, n. 3, p. 035105, 2007.

THOMAS, J.; CHITRA, K.; R. Nanogold Doped TiO2 Nanotubes: Efficient Solar Photocatalyst for the Degradation of Endosulfan. Materials Focus, v. 3, n. 3, p. 233-238, 2014.

TIAN, W. C.; HO, Y. H.; CHEN, C. H.; KUO, C. Y. Sensing performance of precisely ordered TiO2 nanowire gas sensors fabricated by electron-beam lithography. *Sensors, 13,* 865-874, 2013.

WU, X.; JIANG, Q. Z.; MA, Z. F.; FU, M.; & SHANGGUAN, W. Synthesis of titania nanotubes by microwave irradiation. Solid state communications, v. 136, n. 9, p. 513-517, 2005.

YAO, B. D.; CHAN, Y. F.; ZHANG, X. Y.; ZHANG, W. F.; YANG, Z. Y.; & WANG, N. Formation mechanism of TiO2 nanotubes. Applied Physics Letters, v. 82, p. 281-283, 2002.

YU, Y.; ZHANG, P.; GUO, L.; CHEN, Z.; WU, Q.; DING, Y; ... & CAO, Y. The Design of TiO2 Nanostructure (Nanoparticle, Nanotube and Nanosheet) and Their Photocatalytic Activity. The Journal of Physical Chemistry C, 2014.

YUAN, Z. Y.; SU, B. L. Titanium oxide nanotubes, nanofibers and nanowires. Colloids and Surfaces A, v. 241, p. 173-183, 2004.

ZARBIN, A. J. G.; OLIVEIRA, M. M. Carbon nanostructures (nanotubes and graphene): Quo Vadis? Quimica Nova, v. 36, n. 10, p. 1533-1539, 2013. ZHANG, Y.; FU, W.; YANG, H.; QI, Q.; ZENG, Y.; ZHANG, T.; & ZOU, G. Synthesis and characterization of TiO_2 nanotubes for humidity sensing. Applied Surface Science, v. 254, n. 17, p. 5545-5547, 2008.

ZHANG, X. X.; ZHANG, J. B.; JIA, Y. C.; XIAO, P.; TANG, J. TiO2 nanotube array sensor for detecting the SF6 decomposition product so2. *Sensors*, *12*, p. 3302-3313, 2012.

APPENDIX A - PROGRAMS DEVELOPED IN SHELL BASH LANGUAGE ON LINUX TO GENERATE TITANIUM DIOXIDE NANOTUBES AND PLANES

```
#######################################################################

#!/bin/bash

#Programming for the creation of TiO2 pianos and nanotubes - RUTILO #Authors: Gustavo Olinto da Silva and José Divino dos Santos

# ########## External Variables
#######################################################################

np=$1 #Number of 2(TiO2) units in the YZ plane;

npx=$2 #Number of 2(TiO2) units on the X axis;

ot=$6 #Structure optimization option: yes(1) no(0).

# ######### Internal Variables
#######################################################################

a=4. 5937#Network parametre  a of the rutile crystal;

b=4. 5937#Network parameter  b of the rutile crystal;

c=2. 9587#Network parameter  c of the rutile crystal;

pi=' echo "scale=4; 4*a(1)″ | bc  -l'                    #Pi meson value;

te=' echo "scale=4; -1*a(-1*${b}/${a})"|bc -1' #Aligning the piano (110);

dfi=' echo "scale=4; 2*${pi}/${np}"|bc -l' #Variation of the angle;

pe=' echo "scaie=4; ${np}*${c}"|bc -i' #Perimeter;

ra=' echo "scale=4; ${pe}/(2*${pi})"|bc -l' #Ray;

# ######### Growth of the plane in the z-axis and diagonal in the xy-plane
#######################################################################
```

```
dz=' echo "scale=4; ${c}"|bc -l'                    #Growth on the z-axis;

dxy=' echo "scale=4; sqrt^{a}^2 + ${bp2)"|bc -l' #Diagonal growth in the xy plane

#  ######### Translation of crystal atoms

###################################################################

tx=' echo "scale=4; -${a}/2"|bc -l'

ty=' echo "scale=4; -${b}/2"|bc -l'

tz=' echo "scale=4; -${c}/2"|bc -l'

######### Positions of the atoms in the rutile crystal

###################################################################

x1i=' echo "scale=4;(1/4)*${a}"|bc -l '

y1i=' echo "scale=4;(1/4)*${b}"|bc -l '

z1i=' echo "scale=4;(0/4)*${c}"|bc -l '

x2i=' echo "scale=4;(2/4)*${a}"|bc -l '

y2i=' echo "scale=4;(2/4)*${b}"|bc -l '

z2i=' echo "scale=4;(2/4)*${c}"|bc -l '

x3i=' echo "scale=4;(3/4)*${a}"|bc -l '

y3i=' echo "scale=4;(3/4)*${b}"|bc -l '

z3i=' echo "scale=4;(0/4)*${c}"|bc -l '

x4i=' echo "scale=4;(3/4)*${a}"|bc -l '

y4i=' echo "scale=4;(1/4)*${b}"|bc -l '

z4i=' echo "scale=4;(2/4)*${c}"|bc -l '

x5i=' echo "scale=4;(4/4)*${a}"|bc -l '

y5i=' echo "scale=4;(o/4)*${b}"|bc -l '

z5i=' echo "scale=4;(0/4)*${c}"|bc -l '
```

```
x6i=' echo "scale=4;(5/4)*${a}"|bc -l '

y6i=' echo "scale=4;-(1/4)*${b}"|bc -l '

z6i=' echo "scale=4;(2/4)*${c}"|bc -l '

############ Performing plane translation and rotation operations
(110) on the z-axis ########################## 45 degree rotation = arctg(b/a) |
Internal variable ${te} ######################################

x1=' echo "scale=4; (${x1i} + ${tx})*c(${te}) - (${y1i} +
${ty})*s(${te})"|bc -l'

y1=' echo "scale=4; (${x1i} + ${tx})*s(${te}) + (${y1i} +
${ty})*c(${te})"|bc -l'

z1=' echo "scale=4; (${z1i} + ${tz})"|bc -l'

x2=' echo "scale=4; (${x2i} + ${tx})*c(${te}) - (${y2i} + ${ty})*s(${te})"|bc -l'

y2=' echo "scale=4; (S{x2i} + ${tx})*s(${te}) + (${y2i} + ${ty})*c(${te})"|bc -l'

z2=' echo "scale=4; (${z2i} + ${tz})"|bc -l'

x3=' echo "scale=4; (${x3i} + ${tx})*c(${te}) - (${y3i} +
${ty})*s(${te})"|bc -l'

y3=' echo "scale=4; (${x3i} + ${tx})*s(${te}) + (${y3i} + ${ty})*c(${te})"|bc -l'

z3=' echo "scale=4; (S{z3i} + ${tz})"|bc -l'

x4=' echo "scale=4; (S{x4i} + S{tx})*c(S{te}) - (S{y4i} +
${ty})*s(${te})"|bc -l'

y4=' echo "scale=4; (S{x4i} + S{tx})*s(S{te}) + (S{y4i} +
${ty})*c(${te})"|bc -l'

z4=' echo "scale=4; (S{z4i} + ${tz})"|bc -l'

x5=' echo "scale=4; (S{x5i} + S{tx})*c(S{te}) - (S{y5i} +
```

```
${ty})*s(${te})"|bc -l'
y5=' echo "scale=4; (S{x5i} + S{tx})*s(S{te}) + (S{y5i} +
${ty})*c(${te})"|bc -l'
z5=' echo "scale=4; (S{z5i} + ${tz})"|bc -l'
x6=' echo "scale=4; (S{x6i} + S{tx})*c(S{te}) - (S{y6i} +
${ty})*s(${te})"|bc -l'
y6=' echo "scale=4; (S{x6i} + S{tx})*s(S{te}) + (S{y6i} +
${ty})*c(${te})"|bc -l'
z6=' echo "scale=4; (S{z6i} + ${tz})"|bc -l'
#  Conservation of variables
x1s=S{x1}; y1s=S{y1}; z1s=S{z1};
x2s=${x2}; y2s=${y2}; z2s=${z2};
x3s=${x3}; y3s=${y3}; z3s=${z3};
x4s=S{x4}; y4s=S{y4}; z4s=S{z4};
x5s=${x5}; y5s=${y5}; z5s=${z5};
x6s=${x6}; y6s=${y6}; z6s=${z6};
#  ######### CREATION OF TiO2 PLANES (110) RUTILO
##################################################################
name=plan-TiO2-$ {np}-$ {npx}-ot-S{ot}
echo " PM7 PREC SHIFT1000 DEBUG LET T=1999999 + " >
S{name}.dat
echo " DUMP=250000 XYZ PL GEO-OK SINGLET CHARGE=O " >> S{name}.dat
echo " ${name} " >> S{name}.dat
echo " " >> S{name}.dat
```

```
k=1
while [ ${k} -le ${npx} ]
of
i=1
while [ ${i} -le ${np} ]
of
echo "The ${x1} ${ot} ${y1} ${ot} ${z1} ${ot}" >>
S{name}.dat
echo "Ti ${x2} ${ot} ${y2} ${ot} ${z2} ${ot}" >>
${name}.dat
echo "The ${x3} ${ot} ${y3} ${ot} ${z3} ${ot}" >>
${name}.dat
echo "The ${x4} ${ot} ${y4} ${ot} ${z4} ${ot}" >>
${name}.dat
echo "Ti ${x5} ${ot} ${y5} ${ot} ${z5} ${ot}" >>
${name}.dat
echo "The ${x6} ${ot} ${y6} ${ot} ${z6} ${ot}" >> ${name}.dat
i=' expr ${i} + 1 '
z1=' echo "scale=4; ${z1} + ${dz}"|bc -l'
z2=' echo "scale=4; ${z2} + ${dz}"|bc -l'
z3=' echo "scale=4; ${z3} + ${dz}"|bc -l'
z4=' echo "scale=4; ${z4} + ${dz}"|bc -l'
z5=' echo "scale=4; ${z5} + ${dz}"|bc -l'
z6=' echo "scale=4; ${z6} + ${dz}"|bc -l'
```

```
done
#  Retrieving variables
z1=S{z1s};
z2=S{z2s};
z3=${z3s};
z4=${z4s};
z5=${z5s};
z6=S{z6s};
k=' expr $ {k} + 1 '
x1=' echo "scale=4; ${x1} + ${dxy}"|bc -l'
x2=' echo "scale=4; ${x2} + ${dxy}"|bc -l'
x3=' echo "scale=4; ${x3} + ${dxy}"|bc -l'
x4=' echo "scale=4; ${x4} + ${dxy}"|bc -l'
x5=' echo "scale=4; ${x5} + ${dxy}"|bc -l'
x6=' echo "scale=4; ${x6} + ${dxy}"|bc -l'
done
run-mopac2012.bat ${name}
#  Conservation of units;
x1=${x1s}; y1=${y1s}; z1=${z1s};
x2=${x2s}; y2=${y2s}; z2=${z2s};
x3=${x3s}; y3=${y3s}; z3=${z3s};
x4=${x4s}; y4=${y4s}; z4=${z4s};
x5=${x5s}; y5=${y5s}; z5=${z5s};
x6=${x6s}; y6=${y6s}; z6=${z6s};
```

```
# ########## CREATING NANOTUBES FROM
(TiO2)2 PLANE winding########################
# Rotation in X and translation in YZ;
# Angle is now delta = (2pi.z)/(np.c), varies for each z;
ra1=' echo "scale=4;${ra} - ${y3}"|bc -l ' #Minor radius - atomo1 (for
Within the plan (110))
ra3=' echo "scale=4;${ra} + ${y3}"|bc -l ' #Major radius - atomo3 (for
Out of plan (110))
nome1=nanotubo-TiO2-${np}-${npx}-ot-${ot}
echo " PM7 PREC SHIFT1000 DEBUG LET T=1999999 + " >
${name1}.dat
echo " DUMP=250000 XYZ PL GEO-OK SINGLET CHARGE=0 " >>
${name1}.dat
echo " ${name1} " >> ${name1}.dat
echo " " >> ${name1}.dat
w=1
while [ ${w} -le ${npx} ] do
j=1
while [ ${j} -le ${np} ]
of
del=' echo "scale=4; (2*${pi}*${z1})/(${np}*${c})"|bc -l' #Angle del (deltal)
xlr=' echo "scale=4; ${x1}"|bc -l'
ylr=' echo "scale=4; ${ra1}*c(${de1})"|bc -l'
zlr=' echo "scale=4; ${ra1}*s(${de1})"|bc -l'
```

```
de2=' echo "scale=4; (2*${pi}*${z2})/(${np}*${c})"|bc -l' #Angle de2 (delta2)
x2r=' echo "scale=4; ${x2}"|bc -l'
y2r=' echo "scale=4; ${ra}*c(${de2})"|bc -l'
z2r=' echo "scale=4; ${ra}*s(${de2})"|bc -l'
de3=' echo "scale=4; (2*${pi}*${z3})/(${np}*${c})"|bc -l' #Angle de3 (delta3)
x3r=' echo "scale=4; ${x3}"|bc -l'
y3r=' echo "scale=4; ${ra3}*c(${de3})"|bc -l'
z3r=' echo "scale=4; ${ra3}*s(${de3})"|bc -l'
de4=' echo "scale=4; (2*${pi}*${z4})/(${np}*${c})"|bc -l' #Angle de4
x4r=' echo "scale=4; ${x4}"|bc -l'
y4r=' echo "scale=4; ${ra}*c(${de4})"|bc -l'
z4r=' echo "scale=4; ${ra}*s(${de4})"|bc -l'
de5=' echo "scale=4; (2*${pi}*${z5})/(${np}*${c})"|bc -l' #Angle de5 (delta5)
x5r=' echo "scale=4; ${x5}"|bc -l'
y5r=' echo "scale=4; ${ra}*c(${de5})"|bc -l'
z5r=' echo "scale=4; ${ra}*s(${de5})"|bc -l'
de6=' echo "scale=4; (2*${pi}*${z6})/(${np}*${c})"|bc -l' #Angle de6 (delta6)
x6r=' echo "scale=4; ${x6}"|bc -l'
y6r=' echo "scale=4; ${ra}*c(${de6})"|bc -l'
z6r=' echo "scale=4; ${ra}*s(${de6})"|bc -l'
echo "The ${x1r} ${ot} ${y1r} ${ot} ${z1r} ${ot}" >> S{name1}.dat
echo "Ti ${x2r} ${ot} ${y2r} ${ot} ${z2r} ${ot}" >> ${name1}.dat
echo "The ${x3r} ${ot} ${y3r} ${ot} ${z3r} ${ot}" >> ${name1}.dat
echo "The ${x4r} ${ot} ${y4r} ${ot} ${z4r} ${ot}" >> ${name1}.dat
```

```
echo "Ti ${x5r} ${ot} ${y5r} ${ot} ${z5r} ${ot}" >> ${name1}.dat

echo "The ${x6r} ${ot} ${y6r} ${ot} ${z6r} ${ot}" >> ${name1}.dat

j=' expr ${j} + 1 '

z1=' echo "scale=4; ${z1} + ${dz}"|bc -l' z2=' echo "scale=4; ${z2} + ${dz}"|bc -l'
z3=' echo "scale=4; ${z3} + ${dz}"|bc -l' z4=' echo "scale=4; ${z4} + ${dz} "|bc -l'
z5=' echo "scale=4; ${z5} + ${dz} "|bc -l' z6=' echo "scale=4; ${z6} + ${dz} "|bc -l'

done

#Recovery of variable z;

z1=S{z1s};

z2=${z2s};

z3=${z3s};

z4=${z4s};

z5=${z5s};

z6=${z6s};

w=' expr $ {w} + 1 '

x1=' echo "scale=4; ${x1} + ${dxy}"|bc -l' x2=' echo "scale=4; ${x2} + ${dxy}"|bc -
l' x3=' echo "scale=4; ${x3} + ${dxy}"|bc -l' x4=' echo "scale=4; ${x4} + ${dxy}"|bc
-l' x5=' echo "scale=4; ${x5} + ${dxy}"|bc -l' x6=' echo "scale=4; ${x6} + ${dxy}"|bc
-l'

done

run-mopac2012.bat ${name1}

##########################################################################

##########################################################################
```

```
#!/bin/bash

#Programming for the creation of TiO2 pianos and nanotubes - Anatase
```

#Authors: Gustavo Olinto da Silva and José Divino dos Santos

############External Variables

##

np=$1 #Number of 4(TiO2) units in YZ;

npx=$2 #Number of 4(TiO2) units in X;

ot=$6 #Optimize (1) or don't optimize (0);

############Internal Variables

##

a=3. 7845#Net parametera ;

b=3. 7845#Net parameterb ;

c=9. 5143#Redec . parameter

pi=' echo "scale=9; 4*a(1)″ | bc -l' #Pi meson value;

pe=' echo "scale=9; sqrt^{b}^2 + ${cP2)*${np}"|bc -l' #Value of perimeter;

ra=' echo "scale=9; ${pe}/(2*${pi})"|bc -l' #Value of the radius;

te=' echo "scale=9; -a(${b}/${c})"|bc -l' #Value of the rotation angle;

te2=' echo "scale=9; 2*a(1)″ | bc -l' #Value of the angle of the second rotation.

############Growth of the plane on the Z axis and the diagonal of the plane YZ### dyz=' echo "scale=9; sqrt(${bp2 + S{c}^2)″ | bc -1' #Growth along the X axis;

dz=' echo "scale=9; ${a}"|bc -l' #Growth in in Z ############Coordinates of the experimental network parameters of the anatase ## xi1=' echo "scale=9; (0.000026)*${a}"|bc -1 ' yi1=' echo "scale=9; (1.000134)*${b}"|bc -1 '

```
zıl=' echo "scale=9; (0.000010)*S{c}" | bc -1 '
tx=' echo "scale=9; -$ {xil}" be -Γ          #Translation of the atom in x;
ty=' echo "scale=9; -${yil}" be -Γ           #Translation of the atom in y;
tz=' echo "scale=9; -${zi 1}" be -Γ          #Translation of the atom in z;
xi2=' echo "scale=9; (0.000026)*${a}" | bc -1 '
yi2=' echo "scale=9; (0.500068)*$ {b}" | bc -1 '
zı2=' echo "scale=9; (0.051798)*${c}"|bc -1 '
xi3=' echo "scale=9; (0.000026)*${a}" | bc -1 '
yi3=' echo "scale=9; (1.000134)*${b}"|bc -1 '
zi3=' echo "scale=9; (0.206594)*${c}" | bc -1 '
xi4=' echo "scale=9; (0.000026)*$ {a}" | bc -1 '
yi4=' echo "scale=9; (0.500068)*$ {b}" | bc -1 '
zi4=' echo "scale=9; (0.264592)*${c}" | bc -Γ
xi5=' echo "scale=9; (0.500068)*$ {a}" | bc -1 '
yi5=' echo "scale=9; (0.500068)*${b}" | bc -1 '
zı5=' echo "scale=9; (0.283892)*${c}" | bc -1 '
xi6=' echo "scale=9; (0.000026)*${a}" | bc -1 '
yi6=' echo "scale=9; (0.500068)*${b}" | bc -1 '
zi6=' echo "scale=9; (0.477386)*${c}"|bc -1 '
xi7=' echo "scale=9; (0.500068)*${a}"|bc -1 '
yı7=' echo "scale=9; (0.500068)*$ {b}" | bc -1 '
zi7=' echo "scale=9; (0.499985)*${c}" | bc -1 '
xi8=' echo "scale=9; (0.500068)*${a}"|bc -1 '
```

```
yi8=' echo "scale=9; (0.000026)*${b}" | bc -1 '
zi8=' echo "scale=9; (0.522584)*${c}"|bc -1 '
xi9=' echo "scale=9; (0.500068)*${a}"|bc -1 '
yi9=' echo "scale=9; (0.500068)*${b}" | bc -1 '
zi9=' echo "scale=9; (0.715978)*${c}"|bc -1 '
xi10=' echo "scale=9; (0.500068)*${a}" | bc -1 '
yi10=' echo "scale=9; (0.000026)*${b}" | bc -1 '
zi10=' echo "scale=9; (0.735377)*${c}" | bc -1 '
yi11=' echo "scale=9; (0.000026)*${b}"|bc -l '
zi11=' echo "scale=9; (0.948171)*${c}"|bc -l '
xi12=' echo "scale=9; (0.000026)*${a}"|bc -l '
yi12=' echo "scale=9; (0.000026)*${b}"|bc -l '
zi12=' echo "scale=9; (0.793975)*${c}"|bc -l '
######## Translating to origin tx, ty, tz
######## Rotating the plane (101) on the x-axis, te=-arctg(b/c) and then on the
y-axis (te2=90 degrees) ##############################################
dc=0.00000001 #Incrementing
x1t=' echo "scale=9; (${xi1} + ${tx})"|bc -l ' #Translation to origin
y1t=' echo "scale=9; (${yi1} + ${ty})"|bc -l '
z1t=' echo "scale=9; (${zi1} + ${tz})"|bc -l '
x1w=' echo "scale=9; ${x1t}"|bc -l '                    #Rotation in x;
y1w=' echo "scale=9; ${y1t}*c(${te}) - ${z1t}*s(${te})"|bc -l '
z1w=' echo "scale=9; ${y1t}*s(${te}) + ${z1t}*c(${te})"|bc -l '
y1=' echo "scale=9; ${y1w} + ${dc}"|bc -l ' #Y rotation;
```

```
x1=' echo "scale=9; ${x1w}*c(${te2}) + ${z1w}*s(${te2}) + ${dc}"|bc -l '
z1=' echo "scale=9; -${x1w}*s(${te2}) + ${z1w}*c(${te2}) + ${dc}"|bc -l
'
x2t=' echo "scale=9; (${xi2} + ${tx})"|bc -l ' #Translation to origin;
y2t=' echo "scale=9; (${yi2} + ${ty})"|bc -l '
z2t=' echo "scale=9; (${zi2} + ${tz})"|bc -l '
x2w=' echo "scale=9; ${x2t}"|bc -l '                    #Rotation in x;
y2w=' echo "scale=9; ${y2t}*c(${te}) - ${z2t}*s(${te})"|bc -l '
z2w=' echo "scale=9; ${y2t}*s(${te}) + ${z2t}*c(${te})"|bc -l '
y2=' echo "scale=9; ${y2w} + ${dc}"|bc -l ' #Y rotation;
x2=' echo "scale=9; ${x2w}*c(${te2}) + ${z2w}*s(${te2}) + ${dc}"|bc -l '
z2=' echo "scale=9; -${x2w}*s(${te2}) + ${z2w}*c(${te2}) + ${dc}"|bc -l
'
x3t=' echo "scale=9; (${xi3} + ${tx})"|bc -l ' #Translation to origin;
y3t=' echo "scale=9; (${yi3} + ${ty})"|bc -l '
z3t=' echo "scale=9; (S{zi3} + S{tz})″ | bc -l '
x3w=' echo "scale=9; S{x3t}″ | bc -l '             #Rotation in x;
y3w=' echo "scale=9; ${y3t}*c(${te}) - S{z3t}*s(S{te})″ | bc -l ' z3w=' echo "scale=9; S{y3t}*s(S{te}) + S{z3t}*c(S{te})″ | bc -l ' y3=' echo "scale=9; ${y3w} + S{dc}″ | bc -l ' #Y rotation;
x3=' echo "scale=9; S{x3w}*c(S{te2}) + S{z3w}*s(S{te2}) + S{dc}″ | bc -l '
z3=' echo "scale=9; -S{x3w}*s(S{te2}) + S{z3w}*c(S{te2}) + S{dc}″ | bc -l x4t=' echo "scale=9; (S{xi4} + S{tx})″ | bc -l ' #Translation to origin;
y4t=' echo "scale=9; (${yi4} + s{ty})″ | bc -l '
```

z4t=' echo "scale=9; (S{zi4} + S{tz})" | bc -l '

x4w=' echo "scale=9; S{x4t}" | bc -l ' #Rotation in x;

y4w=' echo "scale=9; ${y4t}*c(${te}) - S{z4t}*s(S{te})" | bc -l '

z4w=' echo "scale=9; S{y4t}*s(S{te}) + S{z4t}*c(S{te})" | bc -l ' y4=' echo "scale=9; ${y4w} + S{dc}" | bc -l ' #Y rotation;

x4=' echo "scale=9; ${x4w}*c(${te2}) + S{z4w}*s(S{te2}) + S{dc}" | bc -l '

z4=' echo "scale=9; -S{x4w}*s(S{te2}) + S{z4w}*c(S{te2}) + S{dc}" | bc -l x5t=' echo "scale=9; (S{xi5} + S{tx})" | bc -l ' #Translation to origin;

y5t=' echo "scale=9; (${yi5} + s{ty})" | bc -l '

z5t=' echo "scale=9; (S{zi5} + S{tz})" | bc -l ' x5w=' echo "scale=9; S{x5t}" | bc - l ' #Rotation in x;

y5w=' echo "scale=9; ${y5t}*c(${te}) - S{z5t}*s(S{te})" | bc -l '

z5w=' echo "scale=9; S{y5t}*s(S{te}) + S{z5t}*c(S{te})" | bc -l ' y5=' echo "scale=9; ${y5w} + S{dc}" | bc -l ' #Y rotation;

x5=' echo "scale=9; ${x5w}*c(${te2}) + S{z5w}*s(S{te2}) + S{dc}" | bc -l '

z5=' echo "scale=9; -S{x5w}*s(S{te2}) + S{z5w}*c(S{te2}) + S{dc}" | bc -l x6t=' echo "scale=9; (S{xi6} + S{tx})" | bc -l ' #Translation to origin;

y6t=' echo "scale=9; (${yi6} + s{ty})" | bc -l '

z6t=' echo "scale=9; (S{zi6} + S{tz})" | bc -l '

x6w=' echo "scale=9; S{x6t}" | bc -l ' #Rotation in x;

y6w=' echo "scale=9; ${y6t}*c(${te}) - S{z6t}*s(S{te})" | bc -l '

z6w=' echo "scale=9; S{y6t}*s(S{te}) + S{z6t}*c(S{te})" | bc -l '

y6=' echo "scale=9; ${y6w} + S{dc}" | bc -l ' #Y rotation;

```
x6=' echo "scale=9; ${x6w}*c(${te2}) + S{z6w}*s(S{te2}) + S{dc}" | bc -l'
z6=' echo "scale=9; -S{x6w}*s(S{te2}) + S{z6w}*c(S{te2}) + S{dc}" | bc -l'
x7t=' echo "scale=9; (S{xi7} + S{tx})" | bc -l' #Translation to origin;
y7t=' echo "scale=9; (S{yi7} + S{ty})" | bc -l'
z7t=' echo "scale=9; (S{zi7} + S{tz})" | bc -l'
x7w=' echo "scale=9; S{x7t}" | bc -l'          #Rotation in x;
y7w=' echo "scale=9; ${y7t}*c(${te}) - S{z7t}*s(S{te})" | bc -l'
z7w=' echo "scale=9; S{y7t}*s(S{te}) + S{z7t}*c(S{te})" | bc -l'
y7=' echo "scale=9; ${y7w} + S{dc}" | bc -l' #Y rotation;
x7=' echo "scale=9; S{x7w}*c(S{te2}) + S{z7w}*s(S{te2}) + S{dc}" | bc -l'
z7=' echo "scale=9; -S{x7w}*s(S{te2}) + S{z7w}*c(S{te2}) + S{dc}" | bc -l'
x8t=' echo "scale=9; (S{xi8} + S{tx})" | bc -l' #Translation to origin;
y8t=' echo "scale=9; (S{yi8} + S{ty})" | bc -l'
z8t=' echo "scale=9; (S{zi8} + S{tz})" | bc -l'
x8w=' echo "scale=9; S{x8t}" | bc -l'          #Rotation in x;
y8w=' echo "scale=9; ${y8t}*c(${te}) - S{z8t}*s(S{te})" | bc -l'
z8w=' echo "scale=9; S{y8t}*s(S{te}) + S{z8t}*c(S{te})" | bc -l'
y8=' echo "scale=9; ${y8w} + S{dc}" | bc -l' #Y rotation;
x8=' echo "scale=9; S{x8w}*c(S{te2}) + S{z8w}*s(S{te2}) + S{dc}" | bc -l'
z8=' echo "scale=9; -S{x8w}*s(S{te2}) + S{z8w}*c(S{te2}) + S{dc}" | bc -l'
x9t=' echo "scale=9; (S{xi9} + S{tx})" | bc -l' #Translation to origin;
y9t=' echo "scale=9; (S{yi9} + S{ty})" | bc -l'
```

```
z9t=' echo "scale=9; (S{zi9} + S{tz})″ | bc -l '

x9w=' echo "scale=9; S{x9t}″ | bc -l '              #Rotation in x;

y9w=' echo "scale=9; ${y9t}*c(${te}) - S{z9t}*s(S{te})″ | bc -l '

z9w=' echo "scale=9; S{y9t}*s(S{te}) + S{z9t}*c(S{te})″ | bc -l '

y9=' echo "scale=9; ${y9w} + S{dc}″ | bc -l ' #Y rotation;

x9=' echo "scale=9; S{x9w}*c(S{te2}) + S{z9w}*s(S{te2}) + S{dc}″ | bc -l '

z9=' echo "scale=9; -S{x9w}*s(S{te2}) + S{z9w}*c(S{te2}) + S{dc}″ | bc -l

'

x10t=' echo "scale=9; (S{xi10} + S{tx})″ | bc -l ' #Translation to origin;

y10t=' echo "scale=9; (S{yi10} + S{ty})″ | bc -l '

z10t=' echo "scale=9; (S{zi10} + S{tz})″ | bc -l '

x10w=' echo "scale=9; S{x10t}″ | bc -l '              #Rotation in x;

y10w=' echo "scale=9; S{y10t}*c(S{te}) - S{z10t}*s(S{te})″ | bc -l '

z10w=' echo "scale=9; ${y10t}*s(${te}) + S{z10t}*c(S{te})″ | bc -l '

y10=' echo "scale=9; S{y10w} + S{dc}″ | bc -l ' #Y rotation;

x10=' echo "scale=9; ${x10w}*c(${te2}) + S{z10w}*s(S{te2}) +

S{dc}″ | bc -l '

z10=' echo "scale=9; -S{x10w}*s(S{te2}) + S{z10w}*c(S{te2}) +

S{dc}″ | bc -l '

x11t=' echo "scale=9; (${xi11} + S{tx})″ | bc -l ' #Translation to origin;

y11t=' echo "scale=9; (S{yi11} + S{ty})″ | bc -l '

z11t=' echo "scale=9; (S{zi11} + S{tz})″ | bc -l '
```

```
x11w=' echo "scale=9; S{x11t}″ | bc -l'          #Rotation in x;

y11w=' echo "scale=9; ${y11t}*c(${te}) - S{z11t}*s(S{te})″ | bc -l'

z11w=' echo "scale=9; S{y11t}*s(S{te}) + S{z11t}*c(S{te})″ | bc -l'

y11=' echo "scale=9; S{y11w} + S{dc}″ | bc -l' #Y rotation;

x11=' echo "scale=9; S{x11w}*c(S{te2}) + S{z11w}*s(S{te2}) +

S{dc}″ | bc -l'

z11=' echo "scale=9; -S{x11w}*s(S{te2}) + S{z11w}*c(S{te2}) +

S{dc}″ | bc -l'

x12t=' echo "scale=9; (S{xi12} + S{tx})″ | bc -l' #Translation to origin;

y12t=' echo "scale=9; (S{yi12} + S{ty})″ | bc -l'

z12t=' echo "scale=9; (S{zi12} + S{tz})″ | bc -l'

x12w=' echo "scale=9; S{x12t}″ | bc -l' #Rotation in x;

y12w=' echo "scale=9; ${y12t}*c(${te}) - ${z12t}*s(${te})"|bc -l ' z12w=' echo
"scale=9; ${y12t}*s(${te}) + ${z12t}*c(${te})"|bc -l'

y12=' echo "scale=9; ${y12w} + ${dc}"|bc -l' #Y rotation;

x12=' echo "scale=9; ${x12w}*c(${te2}) + ${z12w}*s(${te2}) +

${dc}"|bc -l'

z12=' echo "scale=9; -${x12w}*s(${te2}) + ${z12w}*c(${te2}) + ${dc}"|bc -l'

#  ########Saves the initial coordinate values

###########################################################################

x1s=${x1}; y1s=${y1}; z1s=${z1};

x2s=${x2}; y2s=${y2}; z2s=${z2};

x3s=${x3}; y3s=${y3}; z3s=${z3};
```

```
x4s=${x4}; y4s=${y4}; z4s=${z4};
x5s=${x5}; y5s=${y5}; z5s=${z5};
xós=${xó}; yós=${yó}; zós=${zó};
x7 s=$ {x7} ; y7 s=$ {y7} ; z7 s=$ {z7} ;
x8s=${x8}; y8s=${y8}; z8s=${z8};
x9s=${x9}; y9s=${y9}; z9s=${z9};
x10s=${x10}; y10s=${y10}; z10s=${z10};
x11s=${x11}; y11s=${y11};z11s=${z11};
x12s=${x12}; y12s=${y12};z12s=${z12};
#   ########Planar generation (TiO2)4 anatase (101)
##################################################################
nome=plano-anatase-TiO2-${np}-${npx}-ot${ot}
npt=' echo "${np}*${npx}*12"|bc '                              ####Header
for molden;
echo " ${npt} " > ${name}.mol
echo "      " >> ${name}.mol
#   ###Header for mopac;
echo " PM7 PREC SHIFT1000 DEBUG LET T=1999999 + " >
${name}.dat
echo " DUMP=250000 XYZ PL GEO-OK SINGLET CHARGE=0 " >>
${name}.dat
echo " ${name} ${npt} atomos " >> ${name}.dat
echo " " >> ${name}.dat
j=1 while [ ${j} -le ${npx} ] do
```

```
i=1 while [ ${i} -le ${np} ] do

echo "Ti ${x1} ${ot} ${y1} ${ot} ${z1} ${ot} " >> ${name}.dat echo "O ${x2} ${ot} ${y2} ${ot} ${z2} ${ot} " >> ${name}.dat echo "O ${x3} ${ot} ${y3} ${ot} ${z3} ${ot} " >> ${name}.dat echo "Ti ${x4} ${ot} ${y4} ${ot} ${z4} ${ot} " >> ${name}.dat echo "O ${x5} ${ot} ${y5} ${ot} ${z5} ${ot} " >> ${name}.dat echo "O ${x6} ${ot} ${y6} ${ot} ${z6} ${ot} " >> ${name}.dat echo "Ti ${x7} ${ot} ${y7} ${ot} ${z7} ${ot} " >> ${name}.dat echo "O ${x8} ${ot} ${y8} ${ot} ${z8} ${ot} " >> ${name}.dat echo "O ${x9} ${ot} ${y9} ${ot} ${z9} ${ot} " >> ${name}.dat echo "The ${x10} ${ot} ${y10} ${ot} ${z10} ${ot} " >> ${name}.dat echo "The ${x11} ${ot} ${y11} ${ot} ${z11} ${ot} " >> ${name}.dat echo "The ${x12} ${ot} ${y12} ${ot} ${z12} ${ot} " >> ${name}.dat

echo "Ti ${x1} ${y1} ${z1} " >> ${name}.mol echo "O ${x2} ${y2} ${z2} " >> ${name}.mol echo "O ${x3} ${y3} ${z3} " >> ${name}.mol echo "Ti ${x4} ${y4} ${z4} " >> ${name}.mol echo "O ${x5} ${y5} ${z5} " >> ${name}.mol echo "O ${x6} ${y6} ${z6} " >> ${name}.mol echo "Ti ${x7} ${y7} ${z7} " >> ${name}.mol echo "O ${x8} ${y8} ${z8} " >> ${name}.mol echo "O ${x9} ${y9} ${z9} " >> ${name}.mol

echo "Ti ${x10} ${y10} ${z10} " >> ${name}.mol

echo "The ${x11} ${y11} ${z11} " >> ${name}.mol

echo "The ${x12} ${y12} ${z12.} " >> ${name}.mol

i=' expr ${i} + 1 '

x1=' echo "scale=9; ${x1} + ${dyz}"|bc -l ' #grows in X, dyz=sqrt(b^2 + c^2)

x2=' echo "scale=9; ${x2} + ${dyz}"|bc -l ' #grows in X, dyz=sqrt(b^2 + c^2)

x3=' echo "scale=9; ${x3} + ${dyz}"|bc -l ' #grows in X, dyz=sqrt(b^2 + c^2)

x4=' echo "scale=9; ${x4} + ${dyz}"|bc -l ' #grows in X, dyz=sqrt(b^2 + c^2)

x5=' echo "scale=9; ${x5} + ${dyz}"|bc -l ' #grows in X, dyz=sqrt(b^2 + c^2)

x6=' echo "scale=9; ${x6} + ${dyz}"|bc -l ' #grows in X, dyz=sqrt(b^2 + c^2)
```

```
x7=' echo "scale=9; ${x7} + ${dyz}"|bc -l ' #grows in X, dyz=sqrt(b^2 + c^2)
x8=' echo "scale=9; ${x8} + ${dyz}"|bc -l ' #grows in X, dyz=sqrt(b^2 + c^2)
x9=' echo "scale=9; ${x9} + ${dyz}"|bc -l ' #grows in X, dyz=sqrt(b^2 + c^2)
x10=' echo "scale=9; ${x10} + ${dyz}"|bc -l ' #grows in X, dyz=sqrt(b^2 + c^2)
x11=' echo "scale=9; ${x11} + ${dyz}"|bc -l ' #grows in X, dyz=sqrt(b^2 + c^2)
x12=' echo "scale=9; ${x12} + ${dyz}"|bc -l ' #grows in X, dyz=sqrt(b^2 + c^2)
done
x1=${x1s}          ;####Retrieves x1,
x2=$ {x2s} ;       ####Recovers x2,
x3=${x3s}          ;####Retrieves x3,
x4=$ {x4s} ;       ####Recovers x4,
x5=${x5s}          ;####Recovers x5,
x6=$ {x6s} ;       ####Recovers x6,
x7=${x7s}          ;####Recovers x7,
x8=$ {x8s} ;       ####Recovers x8,
x9=$ {x9s} ;       ####Recovers x9,
x10=${x10s}; ####Recovers x10,
x11=S{x11s}        ;####Recovers x11,
x12=S{x12s}        ;####Recovers x12,
j=' expr ${j} + 1 '
z1=' echo "scale=9; ${z1} + ${dz}"|bc -l ' #grows in Z, dz=a or dz=b, a=b
z2=' echo "scale=9; ${z2} + ${dz}"|bc -l ' #grows in Z, dz=a or dz=b, a=b
z3=' echo "scale=9; ${z3} + ${dz}"|bc -l ' #grows in Z, dz=a or dz=b, a=b
z4=' echo "scale=9; ${z4} + ${dz}"|bc -l ' #grows in Z, dz=a or dz=b, a=b
```

```
z5=' echo "scale=9; ${z5} + ${dz}"|bc -l ' #grow in Z, dz=a or dz=b, a=b

z6=' echo "scale=9; ${z6} + ${dz}"|bc -l ' #grows in Z, dz=a or dz=b, a=b

z7=' echo "scale=9; ${z7} + ${dz}"|bc -l ' #grows in Z, dz=a or dz=b, a=b

z8=' echo "scale=9; ${z8} + ${dz}"|bc -l ' #grows in Z, dz=a or dz=b, a=b

z9=' echo "scale=9; ${z9} + ${dz}"|bc -l ' #grows in Z, dz=a or dz=b, a=b

z10=' echo "scale=9; ${z10} + ${dz}"|bc -l ' #grows in Z, dz=a or dz=b, a=b

z11=' echo "scale=9; ${z11} + ${dz}"|bc -l ' #grows in Z, dz=a or dz=b, a=b

z12=' echo "scale=9; ${z12} + ${dz}"|bc -l ' #grows in Z, dz=a or dz=b, a=b

done

x1=${x1s}; y1=${y1s}; z1=${z1s};

x2=${x2s}; y2=${y2s}; z2=${z2s};

x3=${x3s}; y3=${y3s}; z3=${z3s};

x4=${x4s}; y4=${y4s}; z4=${z4s};

x5=${x5s}; y5=${y5s}; z5=${z5s};

x6=${x6s}; y6=${y6s}; z6=${z6s};

x7=${x7s}; y7=${y7s}; z7=${z7s};

x8=${x8s}; y8=${y8s}; z8=${z8s};

x9=${x9s}; y9=${y9s}; z9=${z9s};

x10=${x10s}; y10=${y10s}; z10=${z10s};

x11=${x11s}; y11=${y11s}; z11=${z11s};

x12=S{x12s}; y12=S{y12s}; z12=${z12s};

run-mopac2012.bat ${name}

#############Generation of (TiO2)4 anatase nanotube (101)

#################################################################
```

```
#Z-rotation and Z-translation
#Angle is now delta = (2pi.x)/(np.pe), varying for each x;
ra=' echo "scale=9; (${pe})/(2*${pi})"|bc -1' #Base radius - for the
atoms with y=0
ra1=' echo "scale=9;${ra} + ${y1}"|bc -l '
ra2=' echo "scale=9;${ra} + ${y2}"|bc -l '
ra3=' echo "scale=9;${ra} + ${y3}"|bc -l '
ra4=' echo "scale=9;${ra} + ${y4}"|bc -l '
ra5=' echo "scale=9;${ra} + ${y5}"|bc -l '
ra6=' echo "scale=9;${ra} + ${y6}"|bc -l '
ra7=' echo "scale=9;${ra} + ${y7}"|bc -l '
ra8=' echo "scale=9;${ra} + ${y8}"|bc -l '
ra9=' echo "scale=9;${ra} + ${y9}"|bc -l '
ra10=' echo "scale=9;${ra} + ${y10}"|bc -l '
ra11=' echo "scale=9;${ra} + ${y11}"|bc -l '
ra12=' echo "scale=9;${ra} + ${y11}"|bc -l '
name 1 =nanotubo-anatase-TiO2-$ {np }-$ {npx}-ot$ {ot}
npt=' echo "${np}*${npx}*12"|bc                    '####Header for the
molden;
echo " ${npt} " > ${name1}.mol
echo "      " >> ${name1}.mol
####Head of mopac;
echo " PM7 PREC SHIFT1000 DEBUG LET T=1999999 + " > ${name1}.dat
echo " DUMP=250000 XYZ PL GEO-OK SINGLET CHARGE=0 " >>
```

```
${name1}.dat
echo " ${name1} ${npt} atomos " >> ${name1}.dat
echo " " >> ${name1}.dat j=1
while [ ${j} -le ${npx} ] do
i=1
while [ ${i} -le ${np} ] do
del=' echo "scale=9; (2*${pi}*${x1})/(${pe})"|bc -l' #Del angle (deltal) depends on xl
xlr=' echo "scale=9; ${ra1}*c(${de1}) + ${dc}"|bc -l'
ylr=' echo "scale=9; ${ra1}*s(${de1}) + ${dc}"|bc -l'
zlr=' echo "scale=9; ${z1} + ${dc}"|bc -l'
de2=' echo "scale=9; (2*${pi}*${x2})/(${pe})"|bc -l' #Angle de2 (delta2)
x2r=' echo "scale=9; ${ra2}*c(${de2}) + ${dc}"|bc -l'
y2r=' echo "scale=9; ${ra2}*s(${de2}) + ${dc}"|bc -l'
z2r=' echo "scale=9; ${z2} + ${dc}"|bc -l'
de3=' echo "scale=9; (2*${pi}*${x3})/(${pe})"|bc -l' #Angle de3 (delta3)
x3r=' echo "scale=9; ${ra3}*c(${de3}) + ${dc}"|bc -l'
y3r=' echo "scale=9; ${ra3}*s(${de3}) + ${dc}"|bc -l'
z3r=' echo "scale=9; ${z3} + ${dc}"|bc -l'
de4=' echo "scale=9; (2*${pi}*${x4})/(${pe})"|bc -l' #Angle de4 (delta4)
x4r=' echo "scale=9; ${ra4}*c(${de4}) + ${dc}"|bc -l'
y4r=' echo "scale=9; ${ra4}*s(${de4}) + ${dc}"|bc -l'
z4r=' echo "scale=9; ${z4} + ${dc}"|bc -l'
de5=' echo "scale=9; (2*${pi}*${x5})/(${pe})"|bc -l' #Angle de5 (delta5)
x5r=' echo "scale=9; ${ra5}*c(${de5}) + ${dc}"|bc -l'
```

```
y5r=' echo "scale=9; ${ra5}*s(${de5}) + ${dc}"|bc -l'
z5r=' echo "scale=9; ${z5} + ${dc}"|bc -l'
de6=' echo "scale=9; (2*${pi}*${x6})/(${pe})"|bc -l' #Angle de6 (delta6)
x6r=' echo "scale=9; ${ra6}*c(${de6}) + ${dc}"|bc -l'
y6r=' echo "scale=9; ${ra6}*s(${de6}) + ${dc}"|bc -l'
z6r=' echo "scale=9; ${z6} + ${dc}"|bc -l'
de7=' echo "scale=9; (2*${pi}*${x7})/(${pe})"|bc -l' #Angle de7 (delta7)
x7r=' echo "scale=9; ${ra7}*c(${de7}) + ${dc}"|bc -l'
y7r=' echo "scale=9; ${ra7}*s(${de7}) + ${dc}"|bc -l'
z7r=' echo "scale=9; ${z7} + ${dc}"|bc -l'
de8=' echo "scale=9; (2*${pi}*${x8})/(${pe})"|bc -l' #Angle de8 (delta8)
x8r=' echo "scale=9; ${ra8}*c(${de8}) + ${dc}"|bc -l'
y8r=' echo "scale=9; ${ra8}*s(${de8}) + ${dc}"|bc -l'
z8r=' echo "scale=9; ${z8} + ${dc}"|bc -l'
de9=' echo "scale=9; (2*${pi}*${x9})/(${pe})"|bc -l' #Angle de9 (delta9)
x9r=' echo "scale=9; ${ra9}*c(${de9}) + ${dc}"|bc -l'
y9r=' echo "scale=9; ${ra9}*s(${de9}) + ${dc}"|bc -l'
z9r=' echo "scale=9; ${z9} + ${dc}"|bc -l'
de10=' echo "scale=9; (2*${pi}*${x10})/(${pe})"|bc -l' #Angle de10 (delta10)
x10r=' echo "scale=9; ${ra10}*c(${de10}) + ${dc}"|bc -l'
y10r=' echo "scale=9; ${ra10}*s(${de10}) + ${dc}"|bc -l'
z10r=' echo "scale=9; ${z10} + ${dc}"|bc -l'
de11=' echo "scale=9; (2*${pi}*${x11})/(${pe})"|bc -l' #Angle de11 (delta11)
x11r=' echo "scale=9; S{ra11}*c(S{de11}) + ${dc}"|bc -l'
```

y11r=' echo "scale=9; S{ra11}*s(S{de11}) + ${dc}"|bc -l'

z11r=' echo "scale=9; ${z11} + ${dc}"|bc -l'

de12=' echo "scale=9; (2*${pi}*${x12})/(${pe})"|bc -l' #Angle de12 (delta12)

x12r=' echo "scale=9; S{ra12}*c(S{de12}) + ${dc}"|bc -l'

y12r=' echo "scale=9; S{ra12}*s(S{de12}) + ${dc}"|bc -l'

z12r=' echo "scale=9; ${z12} + ${dc}"|bc -l'

echo "Ti ${x1r} ${ot} ${y1r} ${ot} ${z1r} ${ot} " >> S{name1}.dat

echo "O${x2r} ${ot}${y2r}${ot}${z2r}${ot} " >>

${name1}.dat

echo "O${x3r} ${ot}${y3r}${ot}${z3r}${ot} " >>

${name1}.dat

echo "Ti ${x4r} ${ot} ${y4r} ${ot} ${z4r} " >> ${name1}.dat echo "O ${x5r} ${ot} ${y5r} ${ot} ${z5r} ${ot} " >> ${name1}.dat echo "O ${x6r} ${ot} ${y6r} ${ot} ${z6r} ${ot} " >> ${name1}.dat echo "The ${x7r} ${ot} ${y7r} ${ot} ${z7r} ${ot} " >> ${name1}.dat echo "The ${x8r} ${ot} ${y8r} ${ot} ${z8r} ${ot} " >> ${name1}.dat echo "The ${x9r} ${ot} ${y9r} ${ot} ${z9r} ${ot} " >> ${name1}.dat echo "Ti ${x10r} ${ot} ${y10r} ${ot} ${z10r} ${ot} " >> ${name1}.dat echo "O ${x11r} ${ot} ${y11r} ${ot} ${z11r} ${ot} " >> ${name1}.dat echo "O ${x12r} ${ot} ${y12r} ${ot} ${z12r} ${ot} " >> ${name1}.dat

echo "Ti ${x1r} ${y1r} ${z1r} " >> ${name1}.mol echo "O ${x2r} ${y2r} ${z2r} " >> ${name1}.mol echo "O ${x3r} ${y3r} ${z3r} " >> ${name1}.mol echo "Ti ${x4r} ${y4r} ${z4r} " >> ${name1}.mol echo "O ${x5r} ${y5r} ${z5r} " >> ${name1}.mol echo "O ${x6r} ${y6r} ${z6r} " >> ${name1}.mol echo "Ti ${x7r} ${y7r} ${z7r} " >> ${name1}.mol echo "O ${x8r} ${y8r} ${z8r} " >> ${name1}.mol echo "O ${x9r} ${y9r} ${z9r} " >> ${name1}.mol echo "Ti ${x10r} ${y10r} ${z10r} " >> ${name1}.mol echo "O ${x11r} ${y11r} ${z11r} " >> ${name1}.mol echo "O ${x12r} ${y12r} ${z12r} " >> ${name1}.mol i=` expr ${i} + 1 ` x1=` echo "scale=9; ${x1} + ${dyz}"|bc -l` x2=` echo "scale=9; ${x2} + ${dyz}"|bc -l` x3=` echo "scale=9; ${x3} + ${dyz}"|bc -l` x4=` echo "scale=9; ${x4} + ${dyz}"|bc -l` x5=` echo "scale=9; ${x5} + ${dyz}"|bc -l` x6=` echo "scale=9; ${x6} + ${dyz}"|bc -l` x7=` echo "scale=9; ${x7} + ${dyz}"|bc -l` x8=` echo "scale=9; ${x8} + ${dyz}"|bc -l`

x9=` echo "scale=9; ${x9} + ${dyz}"|bc -l` x10=` echo "scale=9; ${x10} + ${dyz}"|bc -l` x11=` echo "scale=9; ${x11} + ${dyz} "|bc -l` x12=` echo "scale=9; ${x12} + ${dyz} "|bc -l` done x1=${x1s}; x2=${x2s}; x3=${x3s}; x4=${x4s}; x5=${x5s}; x6=${x6s}; x7=${x7s}; x8=${x8s}; x9=${x9s}; x10=${x10s}; x11=${x11s}; x12=${x12s}; j=` expr ${j} + 1 ` z1=` echo "scale=9; ${z1} + ${dz}"|bc -l` z2=` echo "scale=9; ${z2} + ${dz}"|bc -l` z3=` echo "scale=9; ${z3} + ${dz}"|bc -l` z4=` echo "scale=9; ${z4} + ${dz}"|bc -l` z5=` echo "scale=9; ${z5} + ${dz}"|bc -l` z6=` echo "scale=9; ${z6} + ${dz} "|bc -l` z7=` echo "scale=9; ${z7} + ${dz} "|bc -l` z8=` echo "scale=9; ${z8} + ${dz}"|bc -l` z9=` echo "scale=9; ${z9} + ${dz}"|bc -l` z10=` echo "scale=9; ${z10} + ${dz}"|bc -l` z11=` echo "scale=9; ${z11} + ${dz}"|bc -l` z12=` echo "scale=9; ${z12} + ${dz}"|bc -l`

done

run-mopac2012.bat ${name1}

APPENDIX B - PAPERS PUBLISHED AT EVENTS

Model: nanotube: TITANIUM DIOXIDE

Silva, G.O.[1], Taft, C.A.[3], Martins, J.B.L.[2], Longo, E.[4], Iannuzzi, J.D.[1]
[1]Universidade Estadual de Goiás, UnUCET, CP 459, Anápolis/GO, Brasil
[2]UnB - Universidade de Brasília, LQC - Brasília/DF Brasil
[3]CBPF - Centro Brasileiro de Pesquisas Físicas, Rio de Janeiro, RJ, Brasil
[4]LIEC - UNESP-Universidade Estadual Paulista - Araraquara/SP, Brasil

Keywords: TiO_2, nanotubes, rutile, HF, DFT 6-31G*, PM7

Introduction

Methodology

Results and Discussion

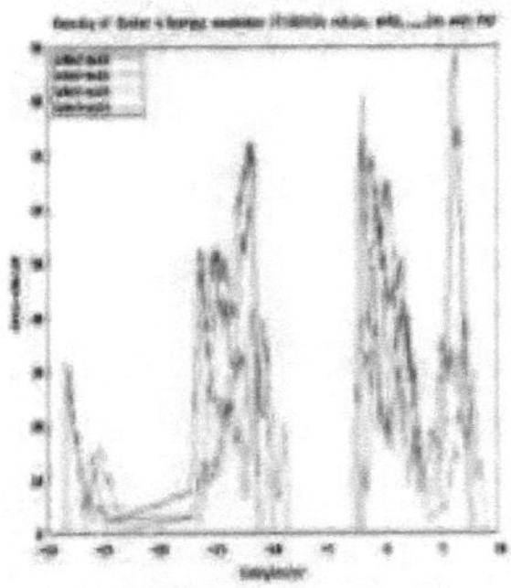

Conclusions

Acknowledgements

References

XIII Encontro da SBPMat
João Pessoa
2014
September 28 to October 02

Simulation of models nanotubes TITANIUM DIOXIDE (rutile and anatase)

Silva, G.O[1], Santos, J.D[1]
[1]Universidade Estadual de Goiás, UnUCET , CP 459, Anápolis/GO, Brasil

gustavo_olinto@hotmail.com

Keywords : TiO_2, nanotubes, anatase, rutile, HF, DFT, 6-31G, 6-31G**, PM7

Nanotubes of the type XY_2 (MnO_2, MgF_2, SnO_2, TiO_2, among others) has been tested in the various forms with the perspective of testing diferent conformations and enable the synthesis of new forms of materials. The calculations used were PM7 semiempirical methods and calculations of type Ab-initio with Hartree-Fock and DFT approach with the basis of the type 6-31G and 6-311G (d, p). This study, the methodology consists of building nanotubes from pre-selected crystal planes. For the rutile crystal selected was the plan [110][1], already the crystal anatase to the selected plan was the plan [101][2]. Using the shell bash, and through of the coordinates of Ti and O atoms and by rotation and translation operations can be simulated a series of nanotubes according to the units TiO_2. The molecular formulas of the nanotubes were defined as $[[(TiO_2)_4]_n]_m$ for nanotubes of anatase and $[[(TiO_2)_2]_n]_m$ for rutile nanotubes, where the coefficients are associated with the diameter and height of the nanotubes. The parameters of interaction distance, HOMO-LUMO, total energy, charges distribution and are a function of n e m, also is assessed the value differences found between ab-initio and semi-empirical methods and their bases. Structures were optimized and the results of the series of nanotubes and plans were collected and there was a tendency for lower valeus energy for nanotubes of higher diameters.

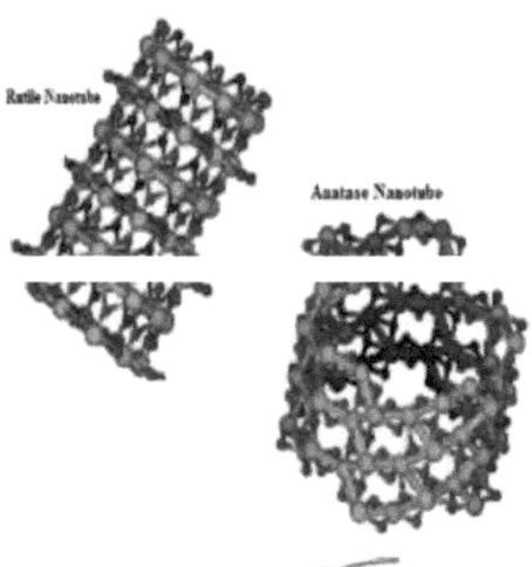

Highlighting the anatase nanotube $[[(TiO_2)_4]_{20}]_3$ (-4.32279 eV), already for structures smaller diameter was verified but higher energy compared, with the anatase nanotube $[[(TiO_2)_4]_3]_1$ (-2.87289 eV). The values found on semi-empirical methods and ab initio HF showed similar results as well as bases 6-311G (d, p) and 6-31G for the same nanotubes. As the nanotube-anatase - $[[(TiO_2)_4]_3]_1$- PM7 (-2.87289 eV) - HF 6-31G (-2.87895 eV) and HF 6-311G (d, p) (-2.87910 eV). The results can provide us with new insights into the use and understanding of these structures, such as carbon nanotubes.

Anatase Nanotubes			
Molecular Formula	ΔE (eV) PM7	ΔE (eV) HF 6-31G	ΔE (eV) HF 6-31G**
{[(TiO2)4]3}1	-2,87289	-2,87895	-2,87910
{[(TiO2)4]5}2	-3,28756	-3,29150	-3,29273
{[(TiO2)4]13}2	-3,84613	-3,84290	-3,84588
{[(TiO2)4]5}3	-3,98576	-3,99572	-3,98477
{[(TiO2)4]20}3	-4,32279	-	-

Rutile Nanotubes			
Molecular Formula	ΔE (eV) PM7	ΔE (eV) HF 6-31G	ΔE (eV) HF 6-31G**
{[(TiO2)2]8}1	-2,21946	-2,23242	-2,24100
{[(TiO2)2]9}2	-2,93157	-2,94127	-2,94342
{[(TiO2)2]14}2	-3,35187	-3,38910	-3,39015
{[(TiO2)2]11}3	-3,87137	-3,86318	-3,86412
{[(TiO2)2]14}3	-4,01651	-3,99857	-3,98461

References

[1] KOWALSKI, P. M.; MEYER, B.; MARX, D. Composition, structure, and stability of the rutile TiO2 (110) surface: Oxygen depletion, hydroxylation, hydrogen migration, and water adsorption. Physical Review B, v. 79, n. 11, p. 115410, 2009.
[2] DIEBOLD, U.; RUZYCKI, N.; HERMAN, G. S.; & SELLONI, A. One step towardsbridging the materials gap: surface studies of TiO2 anatase. Catalysis Today, v. 85, n. 2, p. 93-100, 2003.

MIX
Papier aus verantwortungsvollen Quellen
Paper from responsible sources
FSC® C105338

Printed by Books on Demand GmbH, Norderstedt / Germany